[H.N.S.C. No. 105–18]

THREAT POSED BY ELECTROMAGNETIC PULSE (EMP) TO U.S. MILITARY SYSTEMS AND CIVIL INFRASTRUCTURE

HEARING

BEFORE THE

MILITARY RESEARCH AND DEVELOPMENT SUBCOMMITTEE

OF THE

COMMITTEE ON NATIONAL SECURITY
HOUSE OF REPRESENTATIVES

ONE HUNDRED FIFTH CONGRESS

FIRST SESSION

HEARING HELD
JULY 16, 1997

U.S. GOVERNMENT PRINTING OFFICE

45–989 WASHINGTON : 1998

For sale by the U.S. Government Printing Office
Superintendent of Documents, Congressional Sales Office, Washington, DC 20402
ISBN 0-16-056127-2

45-989 98 - 1

MILITARY RESEARCH AND DEVELOPMENT SUBCOMMITTEE

CURT WELDON, Pennsylvania, *Chairman*

ROSCOE G. BARTLETT, Maryland
JOHN R. KASICH, Ohio
HERBERT H. BATEMAN, Virginia
JOEL HEFLEY, Colorado
JOHN M. McHUGH, New York
JOHN N. HOSTETTLER, Indiana
SAXBY CHAMBLISS, Georgia
VAN HILLEARY, Tennessee
JOE SCARBOROUGH, Florida
WALTER B. JONES, Jr., North Carolina
MICHAEL PAPPAS, New Jersey
BOB RILEY, Alabama
JIM GIBBONS, Nevada

OWEN PICKETT, Virginia
NEIL ABERCROMBIE, Hawaii
MARTIN T. MEEHAN, Massachusetts
JANE HARMAN, California
PAUL McHALE, Pennsylvania
PATRICK J. KENNEDY, Rhode Island
ROD R. BLAGOJEVICH, Illinois
SILVESTRE REYES, Texas
TOM ALLEN, Maine
JIM TURNER, Texas
LORETTA SANCHEZ, California

STEPHEN ANSLEY, *Professional Staff Member*
JEAN REED, *Professional Staff Member*
ROBERT LAUTRUP, *Professional Staff Member*
SUBRATA GHOSHROY, *Professional Staff Member*
TRACY A. WALTER, *Staff Assistant*

CONTENTS

CONTENTS

THREAT POSED BY ELECTROMAGNETIC PULSE (EMP) TO U.S. MILITARY SYSTEMS AND CIVIL INFRASTRUCTURE

HOUSE OF REPRESENTATIVES,
COMMITTEE ON NATIONAL SECURITY,
MILITARY RESEARCH AND DEVELOPMENT SUBCOMMITTEE,
Washington, DC, Wednesday, July 16, 1997.

The subcommittee met, pursuant to call, at 10:20 a.m. in room 2118, Rayburn House Office Building, Hon. Curt Weldon (chairman of the subcommittee) presiding.

OPENING STATEMENT OF HON. CURT WELDON, A REPRESENTATIVE FROM PENNSYLVANIA, CHAIRMAN, MILITARY RESEARCH AND DEVELOPMENT SUBCOMMITTEE

Mr. WELDON. The subcommittee will come to order. Before I get into the subject of today's hearing, I want to apologize to our witnesses and to the public for our lateness. We do have a Republican conference, which is still going on. That is why there is an absence of Republican members. They will be here as soon as the votes are completed for the leadership positions.

Second, I would like to thank the members of the subcommittee and the staff. We just completed our defense authorization bill recently. I was very proud of our subcommittee, primarily because we had pretty much unanimous agreement in the House on the priorities that we established. The best evidence of that was, we had no major amendments relative to the R&D funding levels and issues in the full committee or on the floor of the House, which I think is a testimony to the cooperation and foresight of this subcommittee.

Along that line, we have taken great lengths to create an awareness among the members of our key issues through an aggressive set of hearings where we have had excellent attendance from members on both sides of the aisle; but particularly in the area of missile defense, I think the debate has reached a level that we have not seen, certainly in recent years, in the Congress. In fact, this subcommittee over the past 3 months has sent three separate publications to every Member of the House and Senate trying to raise the level of debate on the issue of missile defense and the threats that are out there.

These documents are public documents and they are available for people that would like to get them, but they are an attempt to continue to have our members be on the cutting edge of information relative to the threats that we have to deal with.

Today's hearing, to some extent, is a follow-on to that effort. Earlier this year, I believe in March, our subcommittee held a hearing,

I think it was the first major congressional hearing on the issue of information warfare. Both in closed and open session, we went into great detail about the threats that are out there on the horizon, and they are here today relative to our information systems. To a large extent, I think what we are going to hear in closed and open session today is a follow-on to that hearing.

Our subcommittee meets to receive testimony today on the effects of electromagnetic pulse, or EMP, on our military systems and civilian infrastructure. I have also invited the members of the Procurement Subcommittee to attend the hearing; and many of them have expressed interest, and I assume, will be here today, committee members who were invited to a June 26 classified briefing by representatives of the Applied Physics Lab at Johns Hopkins on the impact of electromagnetic pulse generated by a high-altitude nuclear blast.

The subcommittee is meeting here today to explore in more detail EMP effects on our military systems and the civilian infrastructure, how confident we are that we can predict these effects, our potential vulnerabilities, what policies and practices guide our efforts to protect our systems, and the steps we have taken and can take to ameliorate these vulnerabilities.

EMP can be generated in several ways, but the widest effects are caused by a high-altitude nuclear blast, although we will ask questions about other than nuclear blasts causing EMP. All of us here understand that the threats posed to our military systems and civilian infrastructure by high-altitude EMP are not new. Atmospheric nuclear tests in the 1950's and 1960's revealed a number of then unanticipated results, including electrical and communications disruptions hundreds of miles from the test sites.

There may be, however, new dimensions in vulnerabilities that we need to look at more carefully. In the 1950's, electronic systems used vacuum tubes, not very sophisticated, but resistant to EMP. Today, computers with more and more microcircuits packed into smaller and smaller chips are key to the efficiency of virtually all commerce in the United States. These have much more computing power, but are also much more sensitive to disruption and more easily disabled by EMP. As our reliance grows, so does our potential vulnerability.

Likewise, potential military vulnerability may be growing. The revolution from military affairs has brought with it a much greater dependence on information technologies. The ability to generate raw data, process it into usable form, and communicate information to the right people and systems is critical to military success, yet the sensors, computers and communications assets essential to this revolution could be vulnerable.

For example, 95 percent of our military communications go through commercial channels. Are we confident that EMP will not disable or disrupt these commercial communications systems? How confident are we that the military could continue to communicate effectively if commercial systems were disrupted or completely disabled by EMP? How thoroughly do we protect our weapons systems from EMP? Are we confident they will continue to function?

At the same time our vulnerabilities may be increasing, the subcommittee is worried about complacency in this area. It is true the cold war is over and the threat of a deliberate nuclear attack by Russia is much lower. That does not necessarily mean we have nothing to worry about. The proliferation of nuclear weapons and ballistic missiles continues. If we come into contact with a rogue nation in possession of just a very few nuclear weapons, perhaps just one, our adversary may or threaten to use one of those weapons to blind our military or to damage our economy or that of an ally.

Yesterday we invited the intelligence community to provide us with their most up-to-date material on EMP threats. We received documents that were 10 years old, interestingly enough, written at that time by a member of our staff, who now is an expert with us, Peter Pry, who at that time was with the agency and the expert on EMP.

We understand that no NIE has been issued on EMP since the 1980's, and this is 1997. We also understand that some of our EMP testing sets have been dismantled or are in disrepair.

I would like to welcome our witnesses today and thank them for taking the time and effort to help us understand the issues. Before I introduce our witnesses, I would add when we get the appropriate number of members, we will vote to close the session following the public statements, especially when our intelligence community is here to provide testimony. At that point in time, we will go into more depth with our intelligence representatives here to talk about the threat analysis.

But that will not come until we have the appropriate number of members of the subcommittee.

To talk about EMP effects, we have Dr. Gary Smith, the director of the Applied Physics Lab from Johns Hopkins University, and Dr. George Ullrich, Deputy Director of Defense Special Weapons Agency. To talk about the impact of these effects on our military systems and commercial infrastructure, we have Gen. Robert T. Marsh, U.S. Air Force, Retired, Chairman of the President's Commission on Critical Infrastructure Protection; the Honorable Gilbert Clinger, Acting Deputy Under Secretary of Defense for Space, and Dr. Lowell Wood of Lawrence Livermore Laboratory.

From the intelligence community we will have Dr. Osias, the National Intelligence Officer for Strategic Systems and Nuclear Proliferation; Dr. Jose Pina from the Central Intelligence Agency; and Dr. Nelson DeGangi from the Defense Intelligence Agency.

Gentleman, we welcome you. Before I open the floor to Dr. Smith, I would ask my friend and colleague, Mr. Pickett to make whatever opening statements he would like to make at this time.

[The prepared statement of Mr. Weldon can be found in the appendix on page 47.]

STATEMENT OF HON. OWEN PICKETT, A REPRESENTATIVE FROM VIRGINIA, RANKING MEMBER, MILITARY RESEARCH AND DEVELOPMENT SUBCOMMITTEE

Mr. PICKETT. Thank you, Mr. Chairman. I wanted also to welcome our witnesses here today.

This is an issue that probably has not gotten as much attention as it should have received. I know it is one that concerns everyone in this room and certainly on this committee. EMP is something that has tremendous adverse potential for our Nation as we move more and more into the electronic age.

I welcome the calling of this hearing by you, Mr. Chairman. I think it is way past the time we should have it. I look forward, as you do, to hearing from the witnesses today and also trying to understand what it is we can do from a policy standpoint to improve our Nation's capabilities to deal with this kind of phenomenon.

Thank you, Mr. Chairman.

Chairman WELDON. Thank you, Mr. Pickett.

With that, we will proceed to our two witnesses.

Dr. Smith, I turn the floor over to you.

STATEMENT OF DR. GARY L. SMITH, DIRECTOR, APPLIED PHYSICS LABORATORY, JOHNS HOPKINS UNIVERSITY

Dr. SMITH. Thank you, Mr. Chairman. Thank you for the opportunity to testify before this subcommittee.

I am Gary Smith. I am the director of the Johns Hopkins University Applied Physics Lab.

Mr. WELDON. Can you move the mike a little closer to you?

Dr. SMITH. Thank you very much. The laboratory that I represent is located in Howard County, MD. Before I begin, Mr. Chairman, I would request permission to insert my full testimony into the record and to confine my remarks this morning to just a brief summary.

The Applied Physics Lab operates under about a dozen task order contracts with a number of major sponsors, covering between 200 and 300 separate tasks in any given year. All of our funding is derived from programs. We have no line-item support. About 80 percent of our funding comes from sponsors within the Department of Defense, and about 15 percent from the National Aeronautics and Space Administration. The rest of the funding comes from various Government departments and agencies.

Recently, as you know, Mr. Chairman, Congressman Bartlett asked me to advise him on the subject of this particular hearing, the electromagnetic pulse produced by a high-altitude nuclear explosion, and the implications for defense systems and capabilities and for the civilian infrastructure. My staff has completed a limited assessment based on a review of the literature, an examination of the basic tenets and interviews with responsible and knowledgeable professionals in the field. I am prepared today to present the results of that assessment.

In short, we have found that the phenomenon is very real, as you have already stated, and is well understood by the nuclear weapons effects community; that our strategic systems and their command, control and communications infrastructure have been designed and built to survive and operate effectively in such an environment; that there would likely be pronounced effects on the civilian infrastructure from such a pulse; that the magnitude and extent of these effects is difficult even to estimate; and that it is probably not feasible to completely protect the entire infrastructure from the effects of such a pulse.

In this testimony, I will first consider electromagnetic pulse—or EMP, as it is called—phenomenology, and I will identify specific EMP-related vulnerabilities for ground system components of the civilian infrastructure. My full testimony discusses protection against EMP as well as nuclear threats to space-based elements of the infrastructure. It specifically reviews threat environments and the effects of prompt and delayed radiation exposure on satellite systems.

Due to the limitations of time this morning, I will not address those aspects in these remarks.

To begin with, electromagnetic energy is really invisible energy traveling in waves which is capable of doing useful work. Such energy exists throughout our environment, and the basic property allows such things as radio and television to work in a useful manner. But electromagnetic energy, even at low levels, can disrupt our lives if we are not careful.

For example, if we put a wristwatch too close to an electric motor, it may cause the watch to become magnetized and run erratically. Everyone also knows that computer floppy disks have to be kept away from magnetic fields or they can be erased or damaged.

Figure 1, which my colleague, Mr. Ron Wiltsie, is illustrating, and which is also on page 2 of your copy, shows the basic phenomenology of an EMP event. The detonation of a nuclear weapon produces high energy gamma radiation that travels radially away from the burst center. When the detonation occurs at high altitudes, greater than about 40 kilometers, the gamma rays directed toward the Earth encounter the atmosphere, where they interact with air molecules to produce positive ions and recoil electrons called Compton electrons, after the man who discovered the effect.

The gamma radiation, interacting with the air molecules, produces charge separation as the Compton recoil electrons are ejected and leave behind the more massive positive ions. The Earth's magnetic field interacts with the Compton recoil electrons and causes charge acceleration, which further radiates electromagnetic energy. EMP is produced by these charge separation and charge acceleration phenomena, which occur in the atmosphere in a layer about 20 kilometers thick and about 30 kilometers above the Earth's surface.

The area of the Earth's surface directly illuminated by EMP is determined entirely by the height of burst. All points on the Earth's surface within the horizon, as seen from the burst point, will experience EMP effects as depicted in figure 2, which is on page 3 of your handout. Note that a burst on the order of 500 kilometers in altitude can cover the entire continental United States.

Mr. WELDON. What strength burst would that be?

Dr. SMITH. It is not terribly burst-strength dependent; almost any burst will produce that kind of radiation. The strength of the field will change at the various radii from the burst point, but it will cover the same area regardless of the strength of the burst.

The amplitude, duration and polarization of the wave depend on the location of the burst, the type of weapon, the yield, and the relative position of the observer. The electric field resulting from a high-altitude nuclear detonation can be on the order of 50 kilovolts per meter with a rise time on the order of 10 nanoseconds and a decay time to half maximum of about 200 nanoseconds. It is very fast.

A localized lightning strike, by comparison, 10 meters away, has a higher peak amplitude by about an order of magnitude, but it rises more slowly than the EMP peak, and therefore it may be simpler to protect against.

It is important to point out, however, that the peak amplitude, signal rise rate, and duration of the EMP wave are not uniform over the illuminated area; the largest peak intensities of the EMP signal occur in that region of the illuminated area where the line of sight to the burst is perpendicular to the Earth's magnetic field. At the edge of the illuminated area, that is, farthest towards the horizon as seen from the burst, the peak field intensity will be about half of the maximum levels, and the EMP fields will be somewhat longer lasting than in the areas where the peak intensities are the largest.

The EMP threat is unique in two respects. First, its peak field amplitude and rise rate are high. These quantities depend upon the rate of rise and the energy of the gamma ray output of the weapon. These features of EMP will induce potentially damaging voltages and currents in unprotected electronic circuits and components.

Second, the area covered by an EMP signal can be immense. As a consequence, large portions of extended power and communications networks, for example, can simultaneously be put at risk. Such far-reaching effects are peculiar to EMP. Neither natural phenomena nor any other nuclear weapon effects are so widespread.

Much of what we depend on today would be susceptible to EMP effects, both in the military and civilian infrastructure. An electromagnetic field interacts with metallic conductors by inducing currents to flow through them. A television antenna, for example, is a collection of metal conductors arranged to facilitate the induced current flow in the frequency range allocated for television broadcasting and to transfer the signal to the receiver.

Other conducting structures, such as aircraft, ships, automobiles, railroad tracks, power lines, and communication lines connected to ground facilities, also effectively serve as receiving antennas for EMP coupling. If the resulting induced currents and voltages, which can be large, are allowed to interact with sensitive electronic circuit and components, they can induce an upset in digital logic circuits or cause damage to the components themselves.

Ground facilities, for example, those housing the large computers central to the functioning of our financial systems, are typically nodes in a larger network and are connected to overhead or buried cables for power and communication. They are also connected to buried pipes for water supply and waste disposal and are typically equipped with communication antennas and distributed security systems of various types. All of these features can direct EMP energy into the facility.

Analyses and simulated EMP testing have shown that currents carried to a facility by long overhead or buried conductors can reach thousands of amperes. Shorter penetrating conductors can carry hundreds of amperes into facilities. Direct EMP penetration through the walls and windows of an unshielded building can induce currents of tens of amperes on illuminated interior conductors.

When EMP energy enters the interior of a potentially vulnerable system, it can cause a variety of adverse effects. These effects include transients, resettable or permanent upset of digital logic circuits, and performance degradation or burnout of electronic components. The collected EMP energy itself can cause malfunction or device failure directly, or it can trigger the system's internal power sources in unintended ways, causing damage by the power sources within the system itself.

In summary, EMP introduces two collectively unique features to the overall picture of system susceptibility to nuclear effects. These features, taken together, distinguish EMP from all other forms, both natural and man-made, of electrical stress and response. First, stresses induced by EMP can significantly exceed those ordinarily encountered in system circuits and components and can thereby increase the probability of upset and burnout occurring in electrical and electronic systems. Second, EMP can cause this increase to occur nearly simultaneously over a large area, about one million square kilometers for a high-altitude burst.

These unique features, together with the lack of occurrence of EMP-like phenomena in the normal day-to-day environment, cause great difficulty in attempting to deal with EMP as a normal engineering problem. In particular, EMP can induce multiple, simultaneous upsets and failures over this wide area.

The coverage and levels that would ensue from an EMP attack are well understood. However, the overall effects on specific terrestrial systems are not as well understood. How much of the telecommunications systems would fail and for how long, how much of the power grid would be disrupted and for how long, how many cars would stop and/or would not start are things that are extremely difficult to predict.

However, just consider what would happen if even a small fraction of the cars on the beltway stopped, and expand that to all the roads throughout the country.

It is also clear that the infrastructure, in general, has become more vulnerable to EMP because of the solid state technology proliferation and the increase in more sensitive components.

I hope that I have been able to give you an idea of the phenomenology associated with EMP and the qualitative effects on our civilian infrastructure. This concludes my statement.

Mr. WELDON. Thank you, Dr. Smith.

[The prepared statement of Dr. Smith can be found in the appendix on page 51.]

Mr. WELDON. Dr. Ullrich, thank you.

STATEMENT OF DR. GEORGE W. ULLRICH, DEPUTY DIRECTOR, DEFENSE SPECIAL WEAPONS AGENCY

Dr. ULLRICH. Thank you, Mr. Chairman. I am George Ullrich, the Deputy Director of the Defense Special Weapons Agency in the Department of Defense. I appreciate the opportunity to appear before you today on this important issue. I would like to briefly summarize my written statement, and I request that my full statement be made part of the record.

Mr. WELDON. Without objection, all of your statements will be made a part of the record.

Dr. ULLRICH. I would be remiss if I did not point out the remarkable coincidence that exactly 52 years ago to the day, on July 16, 1945, the world's first nuclear device was exploded at Trinity site located on an isolated stretch of New Mexico desert in what is now the White Sands Missile Range. Among the team who witnessed that momentous event was Enrico Fermi, Noble Laureate, and perhaps the most brilliant of the Manhattan Project physicists. It has been said of Fermi that he was most likely the last of the 20th century physicists who actually knew all of the physics of his day.

I mention that because it was Enrico Fermi who first, and well in advance of the Trinity event, predicted that a nuclear explosion would generate strong electromagnetic fields and that they would occur over a large spatial domain.

In ensuing years, we have learned a great deal more about nuclear-induced electromagnetic phenomena, and in particular, about the phenomenon of high-altitude electromagnetic pulse, or as we have already referred to it, EMP.

To a layman, a nuclear explosion usually conjures up the image of a mushroom cloud representative of a burst at or near the surface of the Earth. Such a burst creates a variety of immediate effects, most prominently blast and thermal, which fall off rapidly with range, extending up to several miles from ground zero, depending on the yield.

In stark contrast, high-altitude burst, detonated a few hundred kilometers above the surface of the Earth, has as its salient featured effect the ability to simultaneously bathe an entire continent in EMP. The ability of EMP to induce potentially damaging voltages and currents in unprotected electronic circuits and components is well-known. The immense footprint of EMP can therefore simultaneously place at risk unhardened military systems, as well as critical infrastructure systems to include power grids, telecommunication networks, transportation systems, banking systems, medical services, civil emergency systems and so forth.

Another potentially devastating, but less well-known effect of high-altitude nuclear bursts is the artificial pumping of the Van Allen belt with large numbers of electrons. The bomb-induced electrons will remain trapped in these belts for periods exceeding a year and in fact up to several years.

All unhardened satellites in low Earth orbit traversing these enhanced belts can be expected to demise from the total ionizing radiation dose in a matter of days to weeks following one such high-altitude burst. A knowledgeable adversary, armed with a few nuclear weapons, might seek to exploit any such perceived vulnerability, thereby severely degrading the significant U.S. technological advantage built on a foundation of sophisticated electronic systems.

This year's National Security Strategy for a New Century, issued by the White House, warns against the likelihood of an adversary using asymmetric means that avoid our strengths while exploiting our vulnerabilities.

To quote from the report, "Because of our dominance in the conventional military arena, adversaries who challenge the United States are likely to do so using asymmetric means, such as weapons of mass destruction."

The mandate is clear. Hardening systems to the pervasive effects of high-altitude explosions must be part of an overall strategy to balance asymmetries and to disincentivize the acquisition and use of nuclear weapons by potential adversaries.

In keeping with your request, Mr. Chairman, I would next like to provide some additional details regarding the phenomenon of high-altitude EMP, how we protect against it, how we validate the hardness of military systems. I will keep this very brief, since Dr. Smith has already done a superb job on the details.

EMP is highly dependent on the gamma ray output of the weapon, as we have seen. The downward-streaming gamma rays collide with the air molecules, producing high energy electrons in a process called Compton scattering. The Compton electrons, in turn, interact with the Earth's magnetic field, producing an intense, coherent electromagnetic pulse that propagates downward to the surface of the Earth. The EMP effect encompasses an area whose perimeter is defined by the line of sight from the detonation point to the Earth's horizon. Any system within view of the detonation will experience some level of EMP.

For example, if a megaton class weapon were to be detonated 400 kilometers above Omaha, nearly the entire contiguous 48 States would be affected with potentially damaging EMP experience from Boston to Los Angeles, from Chicago to New Orleans.

The frequency range of the pulse is enormously wide, from below 1 hertz to 1 gigahertz, enabling the energy to simultaneously couple to individual electronic components, small components, larger system components, as well as distributed long-line conductors.

One of our earliest experiences with high-altitude EMP dates back to the resumption of atmospheric nuclear testing in 1962 following a 3-year testing moratorium.

During that brief return to atmospheric testing, Starfish Prime, a 1.4 megaton detonation conducted over Johnston Island at an altitude of about 400 kilometers, proved that these effects could have wide-ranging impact on systems. The effects of EMP from the Starfish event were observed in Hawaii, 1,300 kilometers east of the detonation. Street lights and fuses failed on Oahu and telephone service was disrupted on the Island of Kauai.

We have recently learned that Soviet scientists observed similar disruptions following their high-altitude tests. In one test, all protective devices and overhead communication lines were damaged at distances out to 500 kilometers. The same event saw a 1,000 kilometer segment of power line essentially shut down by these effects.

Over the years, we have come to understand how to provide effective protection against the effects of EMP. The basic approach is to envelop the entire system with a integral metallic shield to exclude externally generated electromagnetic fields from the interior. Additionally, all mechanical and electrical penetrations through the shield must be protected.

For example, electrical penetration such as antennas and power connections must be equipped with filters and surge arresters. Windows must be coated with wire mesh or conductive coatings. Doors and utility ports must be sealed with conductive gaskets.

EMP hardening protocols are described in numerous military standards and handbooks. There are user friendly computer codes available to facilitate system hardness design. EMP is well understood.

A particularly good news story is that EMP protection can be quite affordable. If EMP hardening is built in from the start, the cost of EMP hardening is a relatively small fraction of the overall system's cost, approximately 1 to 5 percent. Done after the fact, when the unprotected system has been already fielded, it can be significantly more expensive.

To further explore cost reduction opportunities, my agency has an effort under way to develop integrated hardening techniques that provide protection against multiple hazards. Our initial work focuses on integrated protection against both high-altitude EMP and high-power microwaves.

For EMP testing purposes, the DOD currently operates a suite of simulators capable of large area, threat level field illumination. Also employed are direct current injection techniques and continuous-wave low-level illumination to evaluate shield integrity and energy coupling efficiency. While some EMP simulator facilities have been mothballed over the years, those that remain meet both present and projected customer needs.

To summarize, high-altitude EMP is real and well understood. We know how to harden to it—we know how to test it. High-altitude EMP hardening can be achieved at an affordable cost; I mentioned numbers of 1 to 5 percent.

On a final note, high-altitude EMP does not distinguish between military and civilian systems. Unhardened infrastructure systems, such as commercial power grids, telecommunication networks, as we have discussed before, remain vulnerable to widespread outages and upsets due to high-altitude EMP. While DOD hardens their assets it deems vital, no comparable civilian programs exist. Thus the detonation of one or a few high-altitude nuclear weapons could result in serious problems for the entire U.S. civil and commercial infrastructure.

This is a topic that requires thoughtful attention, and I commend the committee for its interest in it.

Mr. Chairman, that completes my statement. I would be pleased to answer questions.

[The prepared statement of Dr. Ullrich can be found in the appendix on page 69.]

Mr. WELDON. Thank you. We appreciate that statement. If you two wouldn't mind, we are going to bring the other panel up and ask you to join them for questions. That way we will get all the testimony in up front and then begin the questioning, and then after that, get into some technical things we will go closed, and at that point in time, listen to our Intelligence Community.

I would ask our other three witnesses from panel two to come forward. We have Gen. Robert Marsh, U.S. Air Force, Retired, Chairman of the President's Commission on Critical Infrastructure Protection; the Honorable Gilbert I. Klinger, Acting Deputy Under Secretary of Defense for Space; and Dr. Lowell Wood from the Lawrence Livermore Laboratory.

Why don't you come over here with us where there is another——

Mr. ABERCROMBIE. While our witnesses are getting set up there, I observed that test, Mr. Chairman. I was on Oahu.

Mr. WELDON. I didn't think you were born back then.

Mr. ABERCROMBIE. I surely was.

Mr. WELDON. You were a young child, right?

Mr. ABERCROMBIE. Don't I wish. I did want to say, it was the single most sobering experience in my life, because it was, in effect, over Johnston Island, as our witnesses stated, and yet the entire sky lit up in Hawaii, even though it was hundreds of miles away, and there were all these disruptions and so on.

Mr. WELDON. Do you remember the disruptions?

Mr. ABERCROMBIE. That, I don't recall particularly, because everybody was more or less mesmerized by what happened. I think, in the aftermath, there were newspaper observations that were made. But I can tell you that anybody—I know exactly where I was. I know precisely the location on the hill and everywhere when the whole sky lit up. This was, of course, a fairly low-intensity explosion and it made me forever want to participate in some way in seeing that this weapon never got used by anybody for any reason.

Mr. WELDON. I thank the gentleman. Now you have that opportunity to follow through on that goal.

With that, we will turn to our distinguished panelists, and we will start with General Marsh.

STATEMENT OF GEN. ROBERT T. MARSH, U.S. AIR FORCE, RETIRED, AND CHAIRMAN, PRESIDENT'S COMMISSION ON CRITICAL INFRASTRUCTURE PROTECTION

General MARSH. Good morning, Mr. Chairman and members of the subcommittee.

My name is Robert T. Marsh, and I am Chairman of the President's Commission on Critical Infrastructure Protection. I thank you for the opportunity to present my views about the potential of EMP effects as a threat to our critical infrastructures. My perspective arises from my service on the Commission established by Executive Order 13010 on July 15, 1996.

This is a joint Government and industry commission charged with assessing threats to our critical infrastructures and their vulnerabilities. The President identified eight infrastructures as our national life support system. They are: telecommunications, electric power systems, oil and gas transportation and storage, banking and finance, transportation, water supply systems, and emergency services such as medical, police, fire and rescue, and continuity of government services.

The first line of the Executive order says it all: Certain national infrastructures are so vital that their incapacity or destruction would have a debilitating impact on the defense or economic security of the United States.

Our mission is to assess vulnerabilities and threats to the critical infrastructures, identify relevant legal and policy issues and assess how they should be addressed, and recommend to the President a national policy and implementation strategy for protecting critical infrastructures, and, in the process, propose any necessary statutory or regulatory changes.

There are two categories of threats to our infrastructures, physical threats to tangible property and threats of electronic, radio frequency, or computer-based attacks on the information or communications components of critical infrastructures. The Commission's report to the President scheduled for mid-October of this year will propose a national policy and implementation strategy to protect these critical infrastructures from both types of threats and assure their continued operation.

The EMP effects of nuclear weapons, as was noted earlier, were thoroughly studied and well understood during the cold war. At great cost, we hardened our strategic nuclear forces and our critical command and control systems against such effects. We built extensive special test facilities and tested these systems to assure their continued operation under attack.

Obviously, the nuclear threat from hostile nations cannot be dismissed today, but we consider it a remote possibility. Likewise, we consider a terrorist acquiring a nuclear weapon and positioning it at the high altitude necessary for the generation of an EMP burst that would debilitate our infrastructures to be a very remote possibility. Consequently, we are not considering any special measures to counter such a threat, though a high-altitude EMP attack could devastate the telecommunications and other critical infrastructures.

We have also looked at localized radio frequency, or RF, threat innovations and discussed them with RF weapons experts. It is theoretically possible to develop such weapons, but to my knowledge their practicality has not yet been demonstrated. Even if perfected, RF weapons would be targeted at local installations with limited local effect. Nevertheless, progress in this area should be continually and actively assessed. If developments mature as some predict, we will have to address defensive measures.

In summary, there is much promise in this technology, but today I do not see any evidence that suggests capabilities seriously threatening our critical infrastructures.

The present likelihood of a terrorist obtaining a nuclear weapon is uncertain. But even if it happened, generating the high-altitude explosion required to produce a devastating EMP attack would be extremely challenging. There are many easier, less costly, and more dramatic ways for terrorists to use nuclear weapons than delivery to a high altitude. Such an event is so unlikely and difficult to achieve that I do not believe it warrants serious concern at this time. The administration's policy is to prevent proliferation and unauthorized access.

In conclusion, Mr. Chairman, I believe the threat of a major debilitating EMP attack generated by a nuclear weapon is remote at this time. This is also true of the more localized effects of RF weapons, although this area needs to be kept under surveillance and may warrant the development of countermeasures in the future.

This completes my statement, Mr. Chairman. I will be pleased to answer any questions later.

[The prepared statement of General Marsh can be found in the appendix on page 79.]

Mr. WELDON. Thank you, General Marsh.

We will proceed to our next witness.

STATEMENT OF GILBERT I. KLINGER, ACTING DEPUTY UNDER SECRETARY OF DEFENSE FOR SPACE

Mr. KLINGER. Mr. Chairman and members of the committee, it is an honor for me to appear before this committee to address issues related to electromagnetic pulse and national security space systems. As you know, I am Gil Klinger. I am the Acting Deputy Under Secretary for Space with the Office of the Secretary of Defense.

I would note in passing that the Office of the Deputy Under Secretary of Defense for Space is the Secretary of Defense's principal staff assistant for space matters. In that regard, we are developing space policies and monitoring and overseeing the development of space architectures as well as acquisition of all DOD space programs.

Let me direct my opening remarks to EMP as it concerns space systems. Hardening guidance is established indirectly by national and DOD space policy and explicitly by the Joint Staff, and space systems are hardened consistent with that guidance.

In the past, national security space systems have been hardened to varying degrees, and that practice continues today. The fact is that hardening all systems to the maximum possible extent carries with it significant cost implications in light of our assessment of current and projected threats.

We are now studying a range of protection measures which will enhance the entire national security space constellation and its capability against any projected threat. This review will look at a range of alternative protection measures with cost as an independent variable.

The Joint Space Management Board, the senior level DOD and intelligence community body that reviews space matters, recently received and approved a comprehensive space protection study that was conducted by the Department of Defense space architect and the National Reconnaissance Office. Analysis and specific implementation is being planned of the recommendations contained in this study of current and projected issues.

We recognize the growth in the use of space systems by the Department of Defense and, moreover, the criticality of space systems in implementing Joint Vision 2010, the future blueprint for the conduct of U.S. and allied military operations.

In light of this potential massive use of space capabilities, both dedicated national security and supporting commercial space systems will require a review and study of critical infrastructure components within the overall critical infrastructure protection activities currently under way under the direction of General Marsh. For the space systems at issue, DUSD Space is heading this effort.

As Drs. Ullrich and Smith pointed out earlier, the most common usage of the term "electromagnetic pulse" refers to the complex radio frequency wave form generated by a nuclear device being detonated in the atmosphere. This is also known as high-altitude EMP, or HEMP. Spacecraft are far from an atmospheric HEMP event and are not damaged. The effects of HEMP on satellite ground stations, however, can be more disruptive, and those depend on a number of factors such as the altitude of the detonation, the distance of the ground station from the burst itself, and the design of the ground station in terms of the protections that it affords against HEMP effects.

Another form of electromagnetic pulse that was previously discussed is system-generated EMP, or SGEMP. This effect happens only in a vacuum and when x rays produced by a nuclear device strike a satellite and displace electrons throughout the spacecraft. Obviously, the movement of these electrons has detrimental effects for the spacecraft components and its functions. The design and testing of military space systems for vulnerability to SGEMP effects is the responsibility of the individual program offices that are in the business of acquiring those space systems. Obviously, we oversee all facets of those acquisitions.

The adequacy of nuclear survivability features within the overall context of a program acquisition is also considered by the Defense Acquisition Board. For space architectures, these issues are developed by the Department of Defense and NRO architects and are also brought before the joint space management board. Again, as part of our ongoing space protection efforts, we will be looking explicitly at the issue, with its cost kept in mind as an independent variable.

I now would like to spend a few moments discussing radiation-hardened microelectronics. The radiation effects discussed earlier are the result of natural and man-made sources. Natural sources include cosmic rays, charged particles trapped in the Van Allen belts, as well as solar flares which are a feature of solar meteorology.

DOD satellite systems must be designed to survive much more severe radiation environments induced either by hostile actions or by operating in high ambient radiation orbits. For example, the global positioning system is one such system that must tolerate these high levels of naturally occurring radiation. Especially designed and manufactured electronic components used in these systems are termed to be radiant heat hardened or simply rad hard.

In December 1996, an integrated product team reported to the Under Secretary of Defense for Acquisition and Technology on the industrial base that designs and manufactures rad-hard components. The IPT found the following: There is an insufficient commercial business base to support a rad-hard industrial capability. The Government, and primarily DOD, is the principal customer in this niche market. Second, Government investment in advanced radiation hardening technology is adequate at the moment, but there is insufficient funding for transitioning these technologies to wide-scale production.

Finally, Government and industry competency has decreased as the industry downsized, predominantly as a function of the end of the cold war. The IPT's rad-hard strategy was to generate economies of scale by providing the seed money for rad-hard parts development and production, thereby assisting the manufacturers to remain in the business while they develop a non-DOD market inasmuch as commercial space system providers also have an interest in some level of radiation protection.

The IPT specifically recommended: first, establishing a DOD level group to oversee and coordinate DOD investment in radiation hardened electronics with a companion interagency coordination group; second, funding an annual radiation hardened investment program at the $60 million to $70 million level of investment; and, finally, creating a graduate level initiative to train more electrical engineers in radiation hardening technology.

In May 1997 the Under Secretary of Defense approved the IPT's recommendations and directed a DOD-wide initiative implementing them. We will update the committee on the implementation of the DOD rad-hard initiative and the progress in creating an interagency coordination mechanism.

All of these issues become more important as DOD transitions to great dependence upon commercial satellite systems to provide cost-effective and affordable ways of meeting mission requirements. We are studying this issue in conjunction with both U.S. Space Command and the NRO. The vulnerabilities of both national security and supporting commercial assets will have to be taken into account. To the extent we migrate critical space functions to commercial providers, we will need to focus on protection and negotiate appropriate hardness levels as part of the acquisition for these programs.

Mr. Chairman, that completes my statement, and I would be pleased to respond to any questions that you have.

[The prepared statement of Mr. Klinger can be found in the appendix on page 86.]

Mr. WELDON. Thank you very much for your statement.

Dr. Wood.

STATEMENT OF DR. LOWELL WOOD, LAWRENCE LIVERMORE LABORATORY

Dr. WOOD. Thank you, Mr. Chairman.

I appear before you today as an independent technical expert who has worked for the past three decades in both the offensive and defensive aspects of EMP. I am not necessarily representing the positions of the Lawrence Livermore National Laboratories where I work, and I request that my prepared statement be included with my testimony.

Mr. WELDON. With unanimous consent, all of your statements will appear in the record.

Dr. WOOD. Electromagnetic pulses, EMP, generated by high-altitude nuclear explosions have riveted the attention of the military nuclear tactical community for three-and-a-half decades since the first comparatively modest one very unexpectedly turned off the lights over a few million square miles in the mid-Pacific. This EMP also shut down radio stations, turned off cars, burned out telephone systems, and wreaked other mischief throughout the Hawaiian Islands nearly 1,000 miles distant from ground zero.

The potential for even a single high-altitude explosion of a more deliberate character to impose continental-scale devastation of much of the equipment of modern civilization and of modern warfare soon became clear. EMP became a technological substrate for the black humor: Suppose they gave a war and nobody came.

It was EMP-imposed wreckage, at least as much as that due to blast, fire, and fallout, which sobered detail studies of the post-nuclear-attack recovery process. When essentially nothing electrical or electronic could be relied upon to work, even in rural areas far from the blast, it appeared surpassingly difficult to bootstrap American national recovery, and post-attack America in these studies remained stuck in the very early 20th century until electrical equipment and electronic components begin to trickle into a Jeffersonian America from abroad.

For obvious reasons, the entire topic of EMP was highly classified and associated congressional oversight was generally circumspect and conducted entirely in closed session. Indeed, this is the first even partly open session of congressional oversight devoted to the EMP topic which I recall, and I congratulate you, Mr. Chairman, for the extraordinary vision and dedication to bedrock aspects of the Nation's security which this morning's hearings betokens.

The third century following the high-altitude test in the early 1960's saw the expenditure of 5 billion present-day dollars by Defense Special Weapons Agency and its predecessors to develop a detailed working understanding of EMP and its consequences for both our own and our adversaries' military hardware systems. Substantially larger sums were expended by other components of DOD in order to express this understanding as military force in being, primarily to defend especially vital military equipments against EMP's destructive effects. You have heard excellent summaries of these effort from Drs. Smith and Ullrich.

Regrettably, these defensive efforts directed towards strategic military capabilities were not perfectly fruitful. To be sure, there were some outstanding success stories. However, a number of important military systems were quite incompletely defended and some were defended only on paper.

Even more regrettable was the fact that most major military hardware and systems, especially those not considered vital to the conduct of strategic nuclear war, were not hardened against EMP much at all. As a result, at the present time our national profile of vulnerability to EMP attack is highly uneven, with large parts of our military machine and virtually all of the equipment undergirding modern American civilization being utterly EMP vulnerable.

Through the end of the cold war, this posture, though unfortunate, arguably could be tolerated. Only one nation, the Soviet Union, could mount EMP attacks on the United States, and likely only as the first punch of a fight to the death conducted with EMP hardened means.

Indicated responses to any EMP attack then were clear. To be sure, the maximum Soviet capability to impose such attacks still exists in the strategic forces of the Russian Federation, and I unhesitatingly predict that it will continue to exist for many decades to come.

Today, we watch the ongoing diffusion by purchase and by illicit routes, at least as much as by indigenous development, of nuclear weapons technologies throughout the Third World. At the same time, we are compelled to acknowledge the unique opportunities for defeating both advanced U.S. forces abroad and the American Nation itself which are offered to our adversaries by EMP-centered attacks.

You have heard a great deal about the revolution in military affairs and the promise which it justifiably extends for far greater effectiveness of a postrevolutionary American military. You have heard much less about the classic Achilles heel which EMP poses to any information-intensive military force completely dependent for its electronic data flows on EMP-fragile integrated circuits.

There arises the regrettably real prospect that EMP weaponry, assuredly if it is nuclear and perhaps even if nonnuclear, could abruptly transform a future Desert Storm type operation from another historic victory to a memorable American defeat. Such EMP weaponry could also be deployed with only slightly more advanced means from space to rip up the electrical and electronic infrastructure in the American homeland.

Thus, the de facto national policy of nakedness to all of our potentially EMP-armed enemies takes on ever more the character of national scale masochism. It is perverse, irrational, and assuredly not necessary or foreordained. It is therefore most heartening to see the Congress engage this issue for most of corporate DOD has seemingly resigned from it.

With respect, Mr. Chairman, by their recent accomplishments in protecting military and national systems from EMP, not by their undoubtedly splendid policies, their ringing declarations, and their glowing promises for the outyears, should DOD efforts in EMP defense work be known to you and to your subcommittee.

For the sake of America's future in a nuclear multipolar world, I appeal most earnestly to you and your colleagues to remain seized of this vital issue, one of the few which, in and of itself, carries the potential of major military victory or defeat, perhaps even of national well-being or devastation.

Thank you, Mr. Chairman. I will be pleased to respond to questions.

[The prepared statement of Dr. Wood can be found in the appendix on page 94.]

Mr. WELDON. Thank you, Dr. Wood.

If we could bring back our other two panelists from the first panel, we will begin our first round of questions and we will continue with the questioning of these two panels until such time as we get into classified answers, and at that point in time if we do not have a quorum, we will adjourn the public hearing and then I will convene a classified briefing with all of our panelists, including those from the intelligence community. But I would like to keep as much of this in the open for as long as we can, until we get into more sensitive areas.

First of all, I want to thank you all for coming in and appearing. Excuse my ignorance, but up until this year I really didn't know that much about EMP, and it was largely because of the effort of the gentleman to my right and your left, who encouraged me to take a closer look at what EMP—he himself is a physicist and worked at Johns Hopkins—am I correct?

Mr. BARTLETT. Physiologist.

Mr. WELDON. A physiologist, OK—and worked at the Johns Hopkins Center and is constantly pushing us, as are other more technical Members of Congress, to look at things that perhaps we haven't looked at in a thorough enough manner. And we have had a couple of classified briefings.

I would ask our physicists especially on the panel to respond because they are the experts. You say that an EMP burst would occur at an altitude of 250 miles? Is that the appropriate or optimum altitude for an EMP burst to take place? Is that correct, Dr. Ullrich?

Dr. ULLRICH. There is no particular optimum. Of course, it depends on yield. You can get EMP at altitudes ranging from 50 kilometers on up.

Mr. WELDON. Say I wanted to lay down an EMP over the United States. You said in your testimony maybe 250 miles; is that correct?

Dr. ULLRICH. Yes, you would want to be at about 300 or 400 kilometers, 250 miles, and you would be in the megaton class yield. That would probably put a damaging contour, at least the potential for disruption, out to the periphery of the United States.

Mr. WELDON. What would that megaton yieldage range be?

Dr. ULLRICH. Let me just say 1 to 2 megatons.

Mr. WELDON. 1 to 2. Would the other physicists agree with that assessment?

Dr. Wood.

Dr. WOOD. That is certainly a very reasonable estimate. The only remark I would make, in putting that into context, is that component of the Soviet missile force which is believed to be primarily oriented to laying down EMP attacks on the United States had a warhead which was more in the neighborhood of 10 megatons rather than 1, sir.

Mr. WELDON. So 1 to 10 would be a range that——

Dr. WOOD. One is certainly sufficient to do very much the damage as Dr. Ullrich suggested. I would suggest that the 10 megatons was insurance on the part of Soviet planners.

Mr. WELDON. At a 250-mile altitude, that would pretty much cause damage in the United States of the type that you talked about which would include our power systems?

Dr. SMITH. It would put fields that were sufficiently strong to cause potential disruption. The actual disruption is not something that we can estimate with certainty.

Mr. WELDON. Well, would you say with a fairly high degree of certainty that we would have impact in our utility systems and our communications systems?

Dr. SMITH. Yes.

Mr. WELDON. Would you agree?

Dr. WOOD. Sir, we have substantial impact on our communications and utilities systems from solar magnetic storms which impose far, far lower fields on the long transmission lines and communications systems than, you know, are involved in EMP. And so there would undoubtedly be impact. I believe that Dr. Smith and Ullrich would agree with me that it is difficult to forecast in detail exactly what the impact would be at any specific point.

Mr. WELDON. How about automobiles? We mentioned that before because of the use of microprocessors and all. Do we have any studies done on the impact on automobiles and vehicles?

Dr. WOOD. Because of exactly what you said, sir, the fact that modern automobiles are very semiconductor- and microprocessor-intensive, the likelihood that they would suffer catastrophic damage so that they could not operate is much larger than it was back in the sixties when some automobiles, at least in anecdotal reports, were shut down due to the Starfish explosion. These were automobiles in the Hawaiian Islands.

Dr. ULLRICH. If I might add, the fact that all of these systems depend on complex electronic subsystems and components certainly makes them vulnerable. But even commercial products in many cases have a certain amount of protection devices built in. There is the concern of electromagnetic interference, and some of the fixes there are similar to what you would do to protect against EMP.

So I think it is premature to say that every car would go belly up under these kind of field conditions. It is certainly safe to say that there is a threat for upset and damage, but I think it is very difficult to predict system response to these kind of insults.

Mr. WELDON. I think a couple of you have alluded to the fact that this type of EMP burst would not necessarily have to be nuclear generated, that there are, in fact, efforts under way that have used nonnuclear sources; is that correct?

Dr. WOOD. Nonnuclear sources would necessarily involve much smaller aerial coverage, sir, simply because there is so much available energy in a nuclear explosion to convert to electromagnetic form. Systems that were powered, for instance, by conventional explosives would have correspondingly much smaller coverage areas. Damage within those smaller areas could potentially be as severe.

Mr. WELDON. I guess the frustrating thing for me in trying to understand, having been briefed by a number of physicists—and I also consulted with physicists at the University of Pennsylvania and Drexel University privately before this hearing, and I might say that all the physicists that I talked to are in general agreement about the impact of EMP and what it would cause, is our assessment of the potential for this type of thing to occur either at the United States or that would impact the United States, and I would like to get into that, because there I see some contradiction, at least in my own mind, among the statements that have been brought before us today.

Believe me, I know this issue is also getting the attention of the Speaker of the House. In fact, at about 3:45 p.m. today he wants me to bring over as many of you as I can get to brief the Speaker personally on this issue. Roscoe has talked to the Speaker about this, and he is personally concerned about the issue. So it is going to get more and more attention in this Congress, I can assure you.

But let me say, anyone who knows my record, I don't think that the threat of an all-out Russian attack is higher today. I think the concern of an accident is there, and I think what really concerns me are the rogue nations, not necessarily the terrorists, and I don't see that being brought out in the testimony. And, General Marsh, I am going to get into your statement in a moment.

I want to ask you all to respond, as physicists and as policy people. If I am the commander of North Korea and I have one nuclear weapon and that weapon is in the range of 1 to 10 kilotons, which I assume it is, and if I have the capability of a Nodong or Taepodong 2, system which I assume can reach an altitude of 250 miles quite easily, General Marsh—at least that is the testimony that has been give to me—and I want to do something to hurt the United States, I think the weapon of choice is to launch that device in the air and wipe out our smart capability and then dare us to respond, because we haven't killed anyone, we haven't hurt any buildings, and we, being a moral Nation, what is our President going to do? Is he going to set off a nuclear strike against North Korea, when they have not killed one person in this country, but it would devastate our entire infrastructure? That is what concerns me.

And maybe I am reading your statement wrong, but your statement appears to me at least—and correct me and say that I am totally wrong—but I believe it appears to me to dismiss it and say that I believe that the threat of a major attack is remote at this time.

And I compare and contrast it to the statement made—I guess it was the statement made by Dr. Ullrich, where he says, as outlined in the National Security Strategy for a New Century—I am quoting him—the White House, May 1997: Our national military strategy also emphasizes the importance of responding to asymmetries. That is unconventional approaches that avoid or undermine our strengths while exploiting our vulnerabilities.

Well, that to me is what EMP is. Here is a White House document in May of this year coming out and saying this is a major concern for us, and, to quote from the report, it goes on to say, because of our dominance in the conventional military arena, adversaries who challenge the United States are likely to do so using asymmetric means such as weapons of mass destruction, and—to continue to quote—to preserve our technological advantage as DOD develops radiation hardened systems and tests them to assure survivability, end quote. Which leads me to believe that it is being taken seriously by DOD.

And then I go on to read the testimony, which I have done, of retired general—Lt. Gen. Robert Schwitzer, who testified before Congress on June 17, and let me read part of his statement into the record. I quote: To help set the stage, recognize, with experts like a former NSA Director, that we are the most vulnerable Nation on Earth to electronic warfare. This thought is echoed by a former CIA Deputy Director and a former Deputy Attorney General who forecast that we will have an electronic Pearl Harbor if we do not accept a wake-up call. Our vulnerability arises from the fact that we are the most advanced Nation electronically and the greatest user of electricity in the world, end quote.

Now those statements seem to be pretty strong indicators that we need to look at this issue seriously and understand. And the only thing I would say, General Marsh, I would ask to provide for the record the appropriate backup intelligence assessments that support the conclusion that the likelihood of this occurring here is remote.

When we asked the intelligence community to respond and give us their assessment, as I told you, the two documents they sent to us were 10 years old. In fact, they were written by our staffer who worked for the CIA at that time. It indicates to me, in their sending information over in preparation for this hearing, nothing has been done of substance by the intelligence community in a written report form in the last 10 years. At least, they didn't send it to us and we asked for it.

So I would ask you to respond to what appears to me to be a contradiction in the potential severity of the threat of EMP on this Nation.

General MARSH. Well, Mr. Chairman, in response, I am not challenging that EMP would have very devastating effects if employed. So that is not the question. I am saying that the employment is, in our judgment, very remote, and even in the scenario of the North Korean, to put a high-altitude missile of any kind over the central part of this country at 250 miles, to me, seems to be quite an extension of practicality.

Mr. WELDON. And all I would say in response, would you provide for the record in a classified context the intelligence information to back up that statement?

[The information referred to is classified and is retained in the committee files.]

General MARSH. I can attempt to do that, Mr. Chairman. I am not sure that I can get anybody to foreclose any possible scenario, but I believe I can support it being a remote possibility.

Mr. WELDON. But the point I am making, general, is we usually respond to threats based on national intelligence estimates and what the intelligence analysts see as emerging threats around the world. And I have just written, in fact, to the head of the CIA to ask why we don't have an updated NIA on this issue.

But my point is, you have made a pretty strong statement and all I am asking for the record is to give us in a classified setting the appropriate backup intelligence documentation to back up your assertion. That is all.

General MARSH. Yes, sir.

Mr. WELDON. Mr. Pickett.

Mr. PICKETT. Thank you, Mr. Chairman.

We seem to have quite a range of opinions here on this issue, and our role in this is trying, from a policy standpoint, to make sure that appropriate action is taken on behalf of our Nation to do what is realistically doable in the economic and technological environment to protect ourselves against the results of EMP. And I suppose this is a very broad question, but from a policy standpoint, what needs to be done today, in your respective opinions, that is technologically feasible and economically realistic to do to counter the EMP threat to our Nation that is not now being done?

Dr. Smith, we can start with you and go down the line.

Dr. SMITH. At this point in our understanding of the phenomenon and its effects, my recommendation would be to—if, in the judgment of the Congress and the administration, this threat deserves increased attention, the attention should be devoted to tests and analysis to try to understand better what the impact would be before we embark on any kind of corrective measures.

Mr. PICKETT. If I just make sure I understand this, because I am not a scientist and you are probably talking way over my head, but you are saying that we know that there is a certain phenomenon that we can identify but what we don't know is what the actual impact may likely be if, in fact, it develops?

Dr. SMITH. Yes, sir, that is it exactly.

Mr. PICKETT. Thank you.

Dr. Wood.

Dr. WOOD. Sir, the basic point is that essentially all of our conventional military capability and all of our civilian infrastructure is highly vulnerable to EMP damage. The dollar numbers in the civilian infrastructure alone can be conservatively estimated at several trillion dollars' worth of infrastructure which is at risk potentially even from a single pulse—several trillion dollars. So the Congress might properly or appropriately be minded to engage the issue on the basis not only that defeat of our conventional military forces but a very, very profound economic damage to our civilian infrastructure is possible.

And what I would suggest as a particular policy initiative, if there was to be only one, would be the mandating of the assessment, the quantitative assessment, of the EMP vulnerability of major national systems, both civilian and military. Just go out and measure how vulnerable these systems are so that you have numbers to inform the subsequent debate with respect to what should be done about it.

Mr. PICKETT. So you are coming down pretty much where Dr. Smith has come down on what we can do from a policy standpoint?

Dr. WOOD. It is certainly feasible, I believe, from a policy standpoint to mandate the measurement of the levels of vulnerability in both civilian and military systems. Once you know that, then there doesn't have to be this groping in the dark sort of problem that I believe we face at the present time of how much is at risk and how great is the problem involved in fixing it.

We will find that some of our problems, as was implied in Dr. Ullrich's statement, can be fixed very cheaply and very quickly. We will find in other things that we may care about, a lot are quite difficult to fix, and we will then be able to face up to our vulnerabilities and face up to the costs involved in fixing them, but doing so in an informed fashion.

Mr. PICKETT. All right.

Secretary—I am sorry; forgive me; I can only see part of your name.

Mr. KLINGER. I have been called infinitely worse in the Pentagon.

I would emphasize two things before giving you a substantive response. One is that I am not a physicist. I am, in fact, by background a policy person. The second is that my comments will really be focused on the space part of the business of the Department of Defense, not on other defense activities and most assuredly not on the broader civilian infrastructure. So that I would couch the boundaries for the comments I will make.

I think there are a couple of things that have to be kept in mind here from a space perspective. We are a work in progress. We have several hundreds of billions of dollars of ground infrastructure and space systems on orbit right now, tens of billions of dollars' worth of systems that are somewhere in the pipeline.

The design freezes for at least many of these systems occurred in the middle of the cold war. Indeed, as the committee is aware, most of these systems were optimized for a conception, a set of circumstances that now have, at least to some degree, either ceased to exist entirely or whose complexion has changed dramatically.

In other words, as the committee is aware, many of our space systems were optimized to operate in an environment in which, once the nuclear threshold were crossed, it would be crossed willfully and not by accident and crossed on a massive scale, not by a rogue nation, and not by accident.

Therefore, a lot of our space systems, the good news is, are very hard against a lot of these threats; as I mentioned in my opening statements, less so for the ground segments, but again it depends on the specific system.

And yet what has happened? We are in a very difficult stage, which is a transition from operating, building, acquiring, and designing systems that have to make the transition from maintaining compatibility with the systems that are on orbit to adapt to a world that has changed fairly dramatically.

I am not making excuses for those vulnerabilities that exist, and there are some, but the fact of the matter is that the Department, those authorities responsible for policymaking and acquisition with regard to space, are cognizant not only of the threats associated with the EMP phenomenology you heard discussed earlier, but in fact we have to place that discussion in a much broader context, which is bounded by two things. One is very heavy constraints on the amount of money that is available to spend on space systems; second is that EMP is one of a range of potential threats with which our space systems have to contend.

The committee is aware, Mr. Weldon, you made mention earlier about the potential of information warfare threats. That is a whole panoply of threats that we have to contend with, and, quite frankly, from an acquisition standpoint, from a policy standpoint, we are charged with a difficult responsibility, which is doing our best to assess the nature of the threats, how real those threats are, quite apart from their technical feasibility, and then taking a look at the amount of resources that we have to spend against the various threats that we agree are credible and are the most likely or, as Dr. Wood points out, whose downside consequences are so dramatic, so catastrophic, that even at a low level of probability we must do something against them.

Now, I would emphasize, with regard to DOD space systems and with regard to the EMP effects you have seen discussed here, you have heard discussed here, my assessment is that we are in reasonable shape. Our space systems are hardened precisely because many of them were designed to operate and pass nuclear command and control information in the very scenario which was the massive nuclear exchange scenario of the cold war. Other systems, that is not the case because their intended use was not designed to obtain in those circumstances.

And, again, when we talk about the communications segments themselves, as well as the ground systems that support those satellite systems, it depends on the system. But this is neither—I would emphasize from a policy perspective, this is very much not an all or nothing situation; it is very much a situation in which, regardless of the threats that we consider, we gauge what we do against our best assessment, and one can call it—the cynic would say it is a guess.

Hopefully, it is a well-informed guess, based on our best intelligence estimates, the best inputs of our engineers and physicists, what is the most likely threat? What are the downside consequences? How much money is available to address them, given that those same moneys have to be expended to provide on-orbit capabilities? And that presents us with a series of tough choices.

As I mentioned in my opening statement, I believe it is a good news story that the Department and the intelligence community have recognized that we must take a much harder look at the security and protection of all of our space systems against the full range of threats, hence the overall study that was just done by the two architects and for which specific implementations will now be carried out.

In addition, from a policy perspective, we will be developing an explicit DOD space system protection policy, coordinated with General Marsh's activities, coordinated throughout the Department. So there is a lot of activity going on.

I believe that we are taking the prudent set of steps in the national security space area to try and fix those areas that we believe need remedial attention, continue those practices that have made sense up to now, and, in fact, try and cope with the new environment in which we operate.

Mr. PICKETT. Thank you.

General Marsh.

General MARSH. Sir, the catastrophic effects of nuclear weapons are not new to this Nation. We have grappled with them in the past, and we have chosen the path of deterrence, and we, in the face of the tremendous expenditure that would have been required to provide asset defense of the Nation and its critical assets, we have decided best to prevent the occurrence and we have put all of our energies and treasure to that purpose. That worked, and it worked very successfully.

And I recognize that there is the possibility of a rogue nation or, for that matter, a rogue terrorist group attempting to perform an EMP attack. I think that is very, very remote, and I would say if we have the treasure to put to this problem, even though it be very, very remote, we ought to prevent it from happening. We ought to deny the access to those rogues and do everything possible to prevent such a happening, rather than to undertake the almost unthinkable task, in my judgment, of protecting against this very, very severe threat, protecting all of our systems.

And I say that in light of many other vulnerabilities that we need to be concerned about in our systems, and we have vulnerabilities to a full spectrum of threats that need attention, need serious attention by this Nation. I view the EMP threat as being at the uppermost and most remote corner of that threat spectrum.

Mr. PICKETT. You are the chairman of the President's Commission on Critical Infrastructure Protection, General Marsh, and you have heard the statements made by Dr. Smith and Dr. Wood about the policy direction they suggested as a way to begin the congressional involvement in this process, and that is to try to identify and quantify the potential impacts of an EMP attack, if I am correctly paraphrasing you, Dr. Smith and Dr. Wood. Would you concur with that view?

General MARSH. Sir, I haven't thought a lot about that. I don't quarrel with any—in fact, I strongly support any rigorous vulnerability analysis of our systems as they incorporate more and more of advanced electronic technology. I think we have to have a continuing vulnerability assessment under way and point to the type of measures that ought to be undertaken to cope with those vulnerabilities. I would not focus it exclusively on EMP if I had that choice.

Mr. PICKETT. I understand there are other threats, and I noticed that the report of your Commission is scheduled to go to the President in October of this year.

General MARSH. Yes, sir.

Mr. PICKETT. And that report will deal, not exclusively but to some extent, with the EMP threat?

General MARSH. The full range of threats as we see them; Yes, sir.

Mr. PICKETT. Thank you.

And Dr. Ullrich.

Dr. ULLRICH. Mr. Pickett, I believe the Department has acted responsibly in the past in dealing with nuclear survivability issues related to military systems. Many of our systems are hardened to operate in these type of environments. However, most of us have had most familiarity in dealing with this peer adversary threat.

I do believe, as pointed out by the chairman, that these new nuke type scenarios in regional conflicts deserve scrutiny, and I think we need to place renewed emphasis and ask the questions of what systems are really expected to operate in these kinds of environments and assess as to whether or not they can do so.

Likewise with civilian infrastructure, before we undertake any expensive fixes, I think the first requirement here is to do an assessment, is to understand what protection mechanisms that are already built in, how they might play against this new environment, and what additional things might ought to be done. So I think some network analysis would be in the offing before we declare that the sky is falling and before we pronounce that everything will demise.

Mr. PICKETT. Do you want to comment on that?

Dr. WOOD. If I might just very briefly, Mr. Pickett, to follow up on that and to extend slightly my previous recommendation to quantitatively assess what our vulnerabilities are.

I think it would be very valuable to, in the same process, task a red team of experts to look at the results of these vulnerability analyses and see how easily different adversaries of different classes of our country might possibly exploit those vulnerabilities. Is it easy? Is it hard? Is it available to 3 nations or 10 nations or 20 nations? How much would they have to spend? And so forth.

One of the striking things that came out of the 1995 Defense Science Board Summer Study on the Revolution of Military Affairs was that while we Americans might have to spend something of the order of $10 billion a year to bring about the revolution, that very effective counters to American capability that would be gained through the Revolution of Military Affairs could be gained by Third

World major regional powers for perhaps $1 billion a year. That is to say, their leverage in economic terms against us was as large as 10 to 1.

This summer study, in which I participated, also found that one of the most striking responses that a major regional superpower can make to an American force projection was the use of EMP to counter a U.S.-led expeditionary force. It was not among the most remote possibilities that the Defense Science Board considered; it was among the most likely.

Mr. PICKETT. All right.

Then, finally, Mr. Chairman, if you will just allow me one other item here.

Mr. Klinger, you stated in your testimony that, for example, electromagnetic pulse policy and standards are established by the Office of the Assistant to the Secretary of Defense for Nuclear and Chemical and Biological Defense Programs. Is that a person we should hear from on this committee to gain additional insight into what action may be taken in the Defense Department with respect to EMP?

Mr. KLINGER. Sir, I can only answer that and give you a personal opinion about that. When it comes to nuclear matters in general, especially the technical matters associated with nuclear weapons and their effects, that office has the lead responsibility within the Office of the Secretary of Defense. That individual is currently Dr. Harold Smith.

Mr. PICKETT. Thank you very much, gentlemen. And thank you, Mr. Chairman.

Mr. WELDON. Thank you, Mr. Pickett.

And before I turn to our colleague, Mr. Bono, Mr. Pickett, on your line of questioning—well, I will pick that up when we go to our second round.

Mr. Bono, I will turn it over to you now.

Mr. BONO. Thank you, Mr. Chairman. I appreciate you calling this hearing, and I am pleased that we have a chairman who looks, to use your word, at indicators. I think that is the key word. There are so many times, if we look at history, where we haven't paid attention to indicators, and it has been very costly, and we have a chairman that stays on top of that and sometimes displays more knowledge about equipment and things than those that inform us, and I think that is to his credit and makes me feel very comfortable.

I feel the same way he does about looking at indicators, and I think that if we don't pay attention to indicators and follow up, that that is where we will be most vulnerable.

Excuse me if I am elementary on EMP, because I am, and I don't have really any knowledge, so if I can ask just a basic question. You refer to EMP in conjunction with space. Is this a weapon, if you will, that can only be used in a space-type circumstance?

Anyone? Mr. Wood? Mr. Smith?

Dr. SMITH. Well, I think the effect that we have been talking about is largely a space-based effect, yes. And as has been mentioned by other of the people at the table, there are other ways in which electromagnetic pulses can be generated. It is not only required that the explosive be discharged in space, but to have the kind of global reach to have the kind of coverage, geographic coverage, that we are imagining, in fact, it has to be a high-altitude burst.

Mr. BONO. In the situation of a war, if somebody wanted to employ it under those circumstances, like the war in the Middle East, could they pull out EMP and use that as an aggressive weapon or as a defensive weapon to knock out some of the smart stuff that we have?

Dr. SMITH. Ground air bursts also produce electromagnetic pulses, but they are of more limited geographic extent.

Mr. BONO. But that could happen?

Dr. SMITH. Yes, sir.

Mr. BONO. That is a possibility?

Dr. SMITH. Yes, sir.

Mr. BONO. Mr. Wood, did you want to comment?

Dr. WOOD. The particular scenario that you just raised, Mr. Bono, is one of very real concern, because in those circumstances, very modest, very short-range rocketry could be used to loft a nuclear explosive over our forces in a force projection situation and impose preferential EMP damage on our forces before they had engaged adversary forces. So there in a sense, are the exceptions.

The only type of capability that is at all advanced would be a small nuclear explosive. You are not interested in covering an entire continent, but rather than stretching 4,000 kilometers, you might only be interested in EMP damage over 400 kilometers, which is a major theater of operations. And in those circumstances, quite modest nuclear explosives on very modest rockets, Scud-type rockets, would suffice to potentially impose very severe damage—as I said in my opening statement, the damage sufficient to make the difference between American victory and American defeat in the military sense.

There is also the prospect, as Dr. Smith has alluded, to using nonnuclear means.

Mr. BONO. I am sorry?

Dr. WOOD. To using nonnuclear means of generating very strong electromagnetic pulses over much more restricted areas, and those potentially operate not just above the atmosphere but deep down in the atmosphere. They are the type that might even be deployed from small civilian planes.

Mr. WELDON. Would the gentleman yield on that point?

Mr. BONO. Yes.

Mr. WELDON. When we get to the classified session, you can ask about the status of weapons held by adversary, nations including Russia, in this area. Is there anything you all can say on the public record about what we know the Russians do have now?

Dr. SMITH. I am not informed on that, sir.

Dr. WOOD. The information which has been published by the Department of Defense on the Soviet military threat, sir, indicated that there were more than a dozen Soviet SS–18 ICBM's which carried large unitary warheads in the 10 megaton class that were believed to have the primary function and military role of conducting an extremely severe military laydown.

Mr. WELDON. I was talking about battlefield capability.

Mr. BONO. I guess it is nonclassified. We are all aware that Russia is extensively involved in R&D. I was just wondering whether they have included researching or perfecting EMP as part of their R&D program that they are involved in now.

Dr. WOOD. They have known about it for every bit as long as we have, sir, and their basic capabilities, intellectual capabilities, and so forth, are in every sense peer ones relative to we Americans.

The technology base on which they operate is well known to be grossly inferior. On the other hand, because of that inferior technology base in many areas of military technology, they have been inclined to be much more innovative and looking toward breakthrough sort of approaches. These have often come a cropper, but every once in a while they have come up with something that represented a very large advance relative to what we expected.

Mr. KLINGER. I think the only thing I would add: We see a lot of reports which obviously we are monitoring on a fairly intense and close basis, as is the intelligence community, about what capabilities were in various stages of development within the Soviet Union and what is still available and the state of those capabilities. There is a tremendous amount of uncertainty about a lot of these things.

Yes, sir, much of my experience is that much of the physics, as you have heard, is well understood. Our experience base is more limited. Our knowledge of effects is more limited. Our knowledge of what the Soviet general staff may have actually taken from the blackboard to operational, fielded, military capabilities is far more limited, far more limited. And thus far, the promises and advertising, which unfortunately is what we frequently see coming out of the former Soviet republics, at least with regard to space systems, has not been backed up by reality. That is not to say they don't exist, but they have not been backed up.

Mr. BONO. Most of the questions that I want to continue with, I guess, should be classified.

I share the chairman's concern that when we have indicators of this nature, that we do not look at these as closely as we can and employ whatever we have to employ as a defensive mechanism, because I am new to this committee, but prior and still before in committee, I have served on a task force under the chairmanship of Mr. Saxton, and the chairman again used the word "unconventional" warfare. And it seems logical that if they do not have the dollars that we do, and we bring out this big equipment, the B–2's and what-have-you, that if they can find a way to immobilize all of this big stuff, that they are going to focus in that area rather than spend a lot of money on hardware.

And in that committee, we found a tremendous amount of activity that is going on in more of a covert line. And to not accept that as a projection for the future in warfare I think would be a big mistake on our part.

I would hope, as the chairman does, that we would pay far more attention to this unconventional activity and defensive activity that is going on to neutralize the kind of dollars and hardware we are spending.

And I thank you, but I would like to go into this more when we do get into closed session. Thank you, Mr. Chairman.

Mr. WELDON. I thank my colleague, and before turning to Mr. Reyes, I would say, just to set the record straight for our witnesses who may think we are looking at this in a vacuum, it was this subcommittee along with the procurement subcommittee that held a lengthy hearing on the issue of information warfare. We heard testimony from, I believe it was Duane Andrews, among others, who chaired the Defense Science Board on that issue, and his recommendation that we should spend $3 billion of additional resources on information warfare above what the President has asked for.

This subcommittee did begin in a very small way to address that concern by plussing up the request in the information warfare technology area by $88 million in the authorization bill. That is not much, but it is a recognition of the fact that we have to provide more assets than we are currently providing in response to the administration's own analysis.

With that, I will turn to Mr. Reyes.

Mr. REYES. Thank you, Mr. Chairman. I don't have any questions for the public portion of the hearing.

Thank you.

Mr. WELDON. Thank you.

Mr. Bartlett, who really is the instigator of this initiative. And we have also been joined by our other physicist in Congress, Mr. Ehlers, who is not a member of this committee but who we are happy to have with us.

And at the appropriate time, Mr. Ehlers, we will give you time for questions.

Mr. Bartlett.

Mr. BARTLETT. Thank you, Mr. Chairman.

I would first like to start by trying to get a consensus on some of the fundamental concerns. First of all, it is my understanding that EMP is an unavoidable consequence of a nuclear detonation; is that correct?

Dr. WOOD. At high altitudes, sir, is where the effects are particularly striking.

Mr. BARTLETT. Well, the effects are always there, except at a ground burst, the area of incineration is roughly the same size as the area of EMP laydown, and so you are not much concerned about EMP laydown if everything has been incinerated because it is just line of sight.

It is my understanding that about 500 kilometers, a 1-megaton burst would lay down an EMP blanket over the entire contiguous 48 States that at the margins—that is at San Diego and Maine—that we would have about 10-kilovolts of energy, that that gets increasingly higher as you come near the ground zero point, which would be, if it were optimized—it would be over what, northeastern Nebraska would be the detonation site?

Dr. SMITH. Yes, sir, except it is 10 kilovolts per meter field strength.

Mr. BARTLETT. That field strength is adequate to cause what kind of damage in which of our systems? Dr. Wood.

Dr. WOOD. Ten kilovolts per meter is where you begin to see substantial damage in all kinds of unprotected semiconductor systems, sir, both civilian and military. This is not to say that you won't see

45-989 98 - 2

it at much lower field strengths, but I don't know of any major military system that has ever been tested unhardened that hasn't suffered widespread damage at 10 kilovolts per meter of EMP equivalent.

Mr. BARTLETT. If we are looking at our civilian infrastructure, what does this 10 kilovolts per meter do to our power grid, for instance?

Dr. SMITH. I don't think that we can honestly say what the effect will be in any given area. We know that there will be large currents, there will be large voltage gradients induced. These will cause some disruption in power systems, but the extent and the duration of these power outages I don't think we have a good handle on, Congressman.

Dr. WOOD. Sir, this is primarily—the uncertainty Dr. Smith referred to is primarily due to the fact, of course, that we have had no large-scale EMP imposed on power systems. We have, however, had God's EMP, which is associated with very severe solar storms. The modulation of the Earth's magnetic field by interaction of the solar wind during those storms causes the low-frequency, or mimics the low-frequency components of EMP, and those are widely reported. For instance, the Federal Power Commission reports that those have resulted in widespread outages, and the magnitudes of upset that induced those outages are known.

And so there is a little bit of a handle from a little bit of data in the real world that indicates that time varying magnetic fields on scales of tens to hundreds or, for that matter, thousands of kilometers of very low amplitude are sufficient to cause the protective equipment in electric power systems to trip out.

Whether or not that equipment reconnects the power system is dependent on circumstances. But there have been large-scale blackouts associated with especially severe solar storms in this country in the past couple of decades.

Mr. BARTLETT. And what is the magnitude—what is the equivalent energy level in these storms as compared to EMP from the weather?

Dr. WOOD. Very small compared to a megaton burst, sir. The time varying magnetic fields are tiny fractions of egals per second. And these are much, much smaller than are calculated to be associated with megaton bursts over the central United States. They are orders of magnitude less than are calculated to occur.

Mr. BARTLETT. So at very low intensity levels, we have had some meaningful disruption of our power systems from naturally occurring activities; and so one would suspect there would be very widespread impact on our power grid with the detonation of a one megaton weapon, for instance?

Dr. WOOD. Those are the formal documented projections, sir, yes.

Mr. BARTLETT. What about the effect on our communication system? Is it more or less vulnerable than a power grid?

Dr. WOOD. It has become exponentially more over the last decades, sir, as our communications systems have come to depend on integrated circuits of smaller and smaller physical dimensions.

The integrated circuit density that we see at the present time is nearly a factor of 10,000 greater than it was just a quarter century ago, and that basically says that the vulnerability of the individual devices have increased correspondingly. There is reason to believe that the semiconductor-based portions of our communication system, which is to say essentially all of it, would be extremely EMP vulnerable. Measurements done on individual systems certainly support that projection.

Mr. BARTLETT. What date was the Johnston Island detonation?

Dr. ULLRICH. I think it was 1962, in July.

Mr. BARTLETT. To what extent were we using microelectronics then as compared to now?

Dr. ULLRICH. Well, the circuits at that time were substantially different than what we use today and that, in fact, is part of the difficulty in using old data and part of the difficulty in assessing infrastructure systems like power grids.

Power grids can use technologies dating back to the early 1900's, all the way up to the present. It is a question of how updated it is. It is very difficult to make an overall assessment and pin it to a specific number.

Mr. BARTLETT. Is it not true that a vacuum tube is one million or so times less susceptible to EMP effects?

Dr. ULLRICH. Absolutely. Vacuum tubes are inherently hard to these kind of effects.

Mr. BARTLETT. If we had some meaningful disruption of communications in 1962 with the technology then in use in our communications systems, would we not expect today enormously more disruption from a burst of similar size?

Dr. ULLRICH. That is certainly possible. Again, one has to go back to the details of what actually was disrupted and what the circuit mechanism was.

I think there possibly is reason for optimism on the telecommunication grids, with a trend toward fiber-optic. But still, you know, you are dealing with repeater stations that involve power and involve electrical signals. So certainly there is still difficulty even if we took that path.

But I think there is reason for optimism. The power grids, however, are more difficult.

Mr. BARTLETT. I just want to reemphasize what you said, that the fiber-optic in and of themselves are immune to this effect. Wherever you have switching and control and so forth, you now have microelectronics and those are vulnerable. So the fact you are transmitting the signals through a nonsusceptible medium really doesn't do much.

Dr. ULLRICH. The point I was trying to make is there is the potential certainly for hardening these repeater stations, along the lines of the protocols that we have developed for military systems. You can isolate them but if you have too many of them, it gets to be a big cost factor. The fiber itself is, in fact, immune to the effect.

As you pointed out, you reconvert to an electrical signal at these repeater stations. But there is a potential for isolating those vulnerabilities and potentially fixing them. That is my point.

Mr. BARTLETT. Yes, sir.

Dr. WOOD. Dr. Ullrich's point is a very important one. Even back when we depended on copper landlines for our strategic war communications systems, those systems were made and tested under quite realistic circumstances. It would be very hard with EMP and other effects by well-known approaches and techniques.

The most obvious one is burying the switching stations and so on well underground and making sure the portals to the surface were well-designed and well-maintained.

So it is feasible, in principle, to make communications systems, even the national communications infrastructure, quite hard. But the processes and techniques are generally not employed. They add slightly to cost, so it is not represented much in current civilian infrastructure.

It would be relatively straightforward in a lot of cases to harden, however, as Dr. Ullrich has implied. It is technically feasible to do so, probably not at great cost.

Mr. BARTLETT. It is my understanding the rise time of this initial pulse travels at the speed of light, so it is essentially instantaneous over the country. The rise times are in nanoseconds, billionths of a second. Do we, in fact—two questions relative to that. One, do we have the ability to simulate those pulses so that we can, in fact, effectively test?

Dr. SMITH. Yes.

Dr. ULLRICH. There are several EMP simulators that are currently operational and are capable of meeting the frequencies and intensity levels (i.e., peak electric field in volt/meter) that are representative of those pulses.

Mr. BARTLETT. To what intensity level?

Dr. ULLRICH. I don't have the number with me right now. I could provide that for the record.

[The information referred to can be found in the appendix beginning on page 119.]

Mr. BARTLETT. It is my understanding we used to have more test facilities. A major test facility has been shut down?

Dr. ULLRICH. Yes, that is the case.

Dr. SMITH. I think we are getting into classified areas, Congressman.

Mr. BARTLETT. We will proceed with that in the classified area. I need to ask one more question in open session. I would presume that an airplane would function as a pretty effective antenna and in coupling with these forces. Conventionally manufactured airplanes, I would suspect, are inherently susceptible?

General MARSH. Yes, sir.

Mr. BARTLETT. One would expect if there was a 1 megaton burst producing 10 kilovolts per meter of energy U.S. wide, that would affect all planes which are in the air?

Dr. WOOD. Yes.

Mr. BARTLETT. Would you expect them to fall out of the air?

General MARSH. I wouldn't expect it to fall out of the air.

Mr. BARTLETT. If all of their control mechanisms were disabled, what would you expect them to do?

General MARSH. I can't answer precisely, Congressman. I would expect the avionics to be considerably disrupted, severely disrupted; and that would obviously influence the ability to land and—depending on what they had for navigation capability and so on, it would be very, very disruptive.

But you are right. As we tested our advanced airborne command posts, the 747's, to make them EMP hard, we found that we had to make very significant investments in order to do that. So that says to take a commercial airplane and make it EMP hard is a major undertaking. They are soft in that sense.

Mr. BARTLETT. Dr. Wood.

Dr. WOOD. Sir, for reasons of safety having to do with large-scale electrical system failures, most civilian passenger airliners can be flown by hard wire and control cable, physical control cable. So, for instance, landing gear can be lowered by a hand-cranked gear system; and the control surfaces, although they are extremely difficult to operate, can be operated by very strong and determined pilots. This has been demonstrated in emergency circumstances on a number of occasions.

It is probably clear that if this attack occurred at night that most of the planes, most of the civilian airliners in the air, would be lost for obvious reasons. They simply won't be able to land. They won't have landing aids, probably no lights on landing strips and so forth. Those would be lost.

Many that were aloft in a daytime attack could probably be landed with a great deal of determination and good fortune on the part of the crews.

But, as has been remarked, military aircraft that are considered to be EMP hard not only have to have a great deal of work done on them but they have to be tested on a regular basis. The Air Force operates DNA and the Air Force operates major test facilities that can, for instance, encompass even an entire B–52; and the qualification of those systems against EMP hardness is considered a very major one by Strategic Command.

Mr. BONO. Would you yield?

Mr. BARTLETT. And then I have one more comment and will yield totally.

Mr. BONO. In that same context, would that knock out the computers in the—it would knock out all the computers, so wouldn't that affect more than just the landing capabilities?

Dr. WOOD. Yes, sir. It is a reasonable projection that most, if not all, modern computer systems exposed to referenced EMP field levels—which are 50 kilovolts per meter, not just 10—but the very high levels you might see in most of the United States—most modern computer systems ranging from laptops to mainframes would wilt. By wilting, they would at least cease to function. In many cases, they would be burned out. So it would require very major maintenance before they could be restored to operation.

Not just computers in aircraft but computers everywhere, other than in this type of very high integrity metallic enclosures that Dr. Ullrich sketched in his opening statement. Computers in any other enclosure than that type would be compromised, if not destroyed outright.

Mr. BARTLETT. Just one other comment, Mr. Chairman; and then I will yield.

In a large city like Washington, DC, hundreds of thousands of homes on any one day, there is a very, very small percentage of those that burn. Yet none of us would sleep well tonight if we had not paid up our fire insurance premium on our home.

Using this analogy, we are not even close in this country to even considering whether or not we should buy insurance against this kind of potential disaster, let alone coming to the point that we are willing to make the necessary investment and insurance premium payment. I think that kind of analogy is a reasonable one.

General Marsh, there is not much probability your home will burn tonight, but I bet you have a fire insurance policy on your home. There is not a large probability that we are going to have an EMP laid down over our country, but the probability is certainly not zero. I would submit that, in terms of relative probabilities, we are more likely to have that as a Nation than the probability that your home will burn tonight.

We as a Nation need to have the same kind of security, confidence, that comes from you having a fire insurance policy on your home. We need to have the equivalent of that in our Nation, and we are not even talking yet about whether or not we should go out and search for an appropriate policy. I think that is where we are, and I am very pleased that we are having this hearing today.

Mr. WELDON. Thank you.

Ms. Sanchez.

Ms. SANCHEZ. Thank you, Mr. Chairman.

Having read quite a bit on this and having had some of you come in to talk to me about it, I have several questions, but probably best left to—of a sensitive nature that are best left to a closed session.

I will at this point yield over my time and move on, hopefully.

Mr. WELDON. Thank you.

Mr. Pappas.

Mr. PAPPAS. No questions.

Mr. WELDON. Mr. Chambliss.

Mr. CHAMBLISS. Thank you, Mr. Chairman.

Not being a physicist and not knowing much about this, I am here to listen and learn. I appreciate the opportunity to ask questions, but I have none at this time.

Mr. WELDON. Thank you.

We have a distinguished guest with us, the physicist from the Science Committee, Mr. Ehlers. Do you have any questions?

Mr. EHLERS. Thank you, Mr. Chairman. I appreciate the opportunity to ask a few.

I am not a member of this committee, as you know, so if I ask any improper questions, you can—that is what I am asking. I have another meeting I am supposed to be at, so I will not be able to stay for the closed hearing.

I haven't looked at EMP for a long time, and I hope you will forgive me for being behind the times on this, but am I to understand that you are saying that the pulse could be generated by a high altitude explosion of any nuclear weapon? Or would it have to be a specially designed weapon to provide the magnitude of pulse you are talking about?

Dr. WOOD. Sir, any nuclear weapon of the type that is stockpiled anywhere at the present time will generate high altitude EMP simply because some finite fraction of the order of percent of its output will be in the gamma rays that drive the pulse, as Dr. Smith and Dr. Ullrich extremely ably described it in their opening statements.

Specially designed types of nuclear weaponry might put out large fractions of their total yield in the form of gamma rays and thus would be correspondingly more threatening as sources of EMP. Equivalently, very low-yield nuclear weaponry that was specially designed could have the same EMP implications as megaton class ordinary weaponry.

Mr. EHLERS. To get the magnitude you are talking about, 10 kilovolt per meter at the fringes of the country, what size convention weapon would you need for that?

Dr. WOOD. Convention nuclear weapon, one megaton class would impose field strengths of at least 10 kilovolts all over the continental United States. The actual field strengths would be more in the neighborhood of 20,000 to 50,000 volts per meter, not 10. And as the fringing fields.

Mr. EHLERS. As you mentioned earlier, God's impact on EMP was the solar effect. What about lightning? How does that compare in magnitude to something like this?

Dr. WOOD. At very close ranges, sir, at 10-meter distances, you can see electric field strengths on the ground which are comparable to or larger than EMP. But, of course, very few pieces of either military or civilian equipment operate within 10 meters at ground level of a lightning stroke.

Also, the frequency characteristics—we are talking about just magnitude. The characteristics of lightning, it is a much lower frequency than EMP; and the threats that it poses to equipment operation have a rather different character.

Mr. EHLERS. Now, the 10 kilovolt per meter figure, has that been verified experimentally? Or is that based on your calculations?

Dr. WOOD. No, sir. Those field strengths were measured in the Pacific tests. The measurements were sporadic. They weren't nearly of as detailed a nature as people would certainly have liked to have had even a few years afterwards, but the advent of comprehensive test ban precluded future measurements.

The calculations of models have been built up by special weapons. The Defense Special Weapons Agency and its predecessors over the last three decades have reproduced the Pacific high altitude test phenomena very well; and these modeling capabilities have also been well exercised and validated in underground nuclear tests, some of a very realistic nature, which have been conducted since then.

Mr. EHLERS. At what altitude were those tests conducted?

Dr. WOOD. The Pacific tests, sir, were altitudes of 100 to several hundred kilometers. The simulated altitudes and underground nuclear experiment circumstances varied from levels of a few dozen kilometers all the way out to very high altitude equivalents, well in excess of 1,000 kilometers.

Mr. EHLERS. Is there something unique about the higher altitudes that enhances the effect or is it simply you are high enough to have a line of sight to a larger region?

Dr. WOOD. The latter.

Mr. EHLERS. Which countries would be capable of doing something of this sort?

Dr. WOOD. Any country that potentially owns a single World War II level nuclear explosive, sir.

As Dr. Ullrich pointed out in his opening statement, the EMP effects are not strongly dependent on the yield of a nuclear explosive. While megaton class definitely delivers larger output than kiloton class, any country that owns fission nuclear explosive weaponry potentially has these capabilities. It is necessary, of course, to get the explosive device to altitudes of 100 to a few hundred kilometers in order to get really widespread effects in order to be able to cover large regions.

But if you are concerned, for instance, with a military theater of operations somewhere in the Third World, you don't need to go to 5,500 kilometers or 300 kilometers. You can get the effects you are interested in very satisfactorily by detonating your device at a 100 kilometer altitude. Because you are interested in, for instance, attacking American forces without doing a lot of EMP damage to your own territory.

Mr. KLINGER. I need to emphasis at this point, from a space perspective, acquisition of the capability to detonate a nuclear device is one thing. Getting it to the altitudes, even 100 kilometers we are talking about here, much less higher, is a different matter altogether.

We are reminded every time NASA or the Department of Defense launches a satellite that, to a great extent, this is an art as much or more than it is a science. It has taken us 30-some-odd years to get to the point where we are the preeminent space-bearing country in the world. Yet, as we saw earlier this year when we tried to launch GPS satellites using the Delta rocket, which has an unparalleled record of success, we had a detonation—we had a malfunction detonation and the loss of the satellites and nearly had significant damage to the launch pad.

Launching rockets into space is not a trivial matter technologically and goes directly to the issues that some of us were discussing earlier with regard to how we judge the threat associated with a high altitude EMP burst and the threat that that poses, independent of the downside consequences.

Dr. WOOD. With respect—the ability to launch rockets into space was one which was extremely massively demonstrated by the Nazis in World War II. The V–2 rocket, which they launched, flew into space over 2,000 times in the process of targeting allied forces in England.

The Soviets took the V–2 engineering crew to the Soviet Union, and they started cranking out in the early 1950's several generations of rockets which are now known to the world as Scuds. Those rockets have been sold very extensively throughout the Third World to the point that even, if you will, a rinky-dink power like North Korea is capable of indigenously producing rockets which are imminently satisfactory for launching not only conventional payloads but nuclear trajectories that go into space.

The Scud rockets used by the Iraqis very effectively, as terror weapons in Operation Desert Storm, flew to altitudes of 150 kilometers, which is imminently satisfactory for the type of regional EMP laydowns I have been referring to.

The Iraqis succeeded in launching not only 80 rockets—79 rockets that landed on or near targets, but they succeeded in launching them at extremely high rates with a very high rate of success. Over 90 percent of all the launches they attempted were successful.

They succeeded, in one particularly striking set of launches, in launching seven of them nearly simultaneously within a time window of 10 minutes. Each of these rockets flew into space. They achieved apogee altitudes of about 150 kilometers. One of these rockets could have carried a nuclear explosive of a kiloton class yield into space and, in particular, detonated it over our forces.

So while I can agree with Secretary Klinger that the very high-tech time of launches that we attempt to do in the United States, particularly launches of spacecraft that survive for long periods of time in deep space and so forth, are still fraught with peril, unfortunately, the Soviet-developed and proliferated capability for Scud-type launches is extremely widely proliferated through the Third World at the present time. Scud-type rockets exist in copy to the extent of over 15,000 Scud class rockets owned by over 30 nations in the world at the present time. So getting to the threshold of space and carrying a nuclear explosive there is something that, unfortunately, is a regrettably potentially widespread—maybe actually widespread capability.

I remind you that, in August of 1995, the director of the Iraqi special weapons projects defected to Jordan and stated—he happened to be Saddam Hussein's senior son-in-law, since deceased—but he stated in a press conference in Amman, Jordan, in August 1995, that his nation had been on track to have Scud-deliverable nuclear weapons by April 1991.

We went to war with Iraq in January, 1991, and finished our business with them in February as far as hostilities were concerned. His program was on track, he said, to produce more than one Scud-deliverable nuclear warhead 2 months after Desert Storm was completed.

That was, I respectfully suggest, a very close call. They went to war too early, in other words. But he stated that his nation was on track to have the ability to deliver nuclear warheads by Scuds in the spring of that year.

So these are things that we have to keep in mind, not just as very remote possibilities but as things that, in the distinctly foreseeable future, we may have to actually face.

Mr. KLINGER. In fact, to add, I would keep a number of other things in mind then.

A number of those Scuds broke up in midair. It is one thing to launch a terror weapon, one that has a nuclear front end on it. Presumably, the party launching it has a vested interest in not only launching it but making sure it detonates not over their own territory but reliably over their intended target.

The point being, this is not nearly as simple a matter as some would imply. Further—I fall back to my previous statement—it is one thing to have a delivery capability; it is another thing to be able to demonstrate the capability, a real, life-threatening capability to detonate a high-altitude nuclear burst over the United States. There is no question that this is a threat over which we are concerned, but the essence of the importance of part of this issue is perspective.

Mr. BONO. If the gentleman will yield.

Dr. Wood, I guess what you are saying is it—there is a little contradiction here in that——

Dr. WOOD. There is a difference of an opinion.

Mr. BONO [continuing]. A difference of opinion that we really don't have to—it really doesn't require as much sophistication, as I understand was being said—as what Mr. Klinger is stating.

Compounding that, wasn't a Scud fired from a truck that just traveled wherever it wanted to go? When it wanted to hide, then it just fired and did a considerable amount of damage. Would that be the case in this situation with EMP?

Dr. WOOD. Scuds are almost invariably fired from mobile launchers, sir. That, of course, is the thing that made it so difficult for us to eliminate the Iraqi Scud-launching capability, is because they tended to pop out of hiding, erect, launch, and pop back into hiding before our strike fighters could localize them and kill them. We didn't have any confirmed kills of Scud-launching capability during Operation Desert Storm, in spite of having flown 8,000 sorties that were specific against Scud launchers.

The launching capability, however, is perhaps not the one that Secretary Klinger was keen on. I don't mean to put words in his mouth, but he was pointing out the obvious concern that—do you want to take your first nuclear weapon you have ever got, stack it on top of a Scud and launch it and hope your firing circuitry and everything else and so forth works and works the very first time out and works without a lot of prior testing? So do you want to really do that?

Hey, that is certainly not something I would ever recommend to my Government, but my Government doesn't own Scuds, and it fired its first nuclear weapon more than a half century ago and so forth.

The Iraqis, obviously, were much less risk averse than we were. They took a lot of chances. If they had understood clearly they had an opportunity to deal a very severe blow to our forces if this particular endeavor succeeded and they had not lost very much if it did not succeed, I would not want to advise my Government that such a step would seem unlikely. It just seems rash to suggest other than be worried about it.

Mr. BONO. In conclusion, in your view then, the potential to launch EMP from a Scud mobile unit is very possible, correct?

Dr. WOOD. Of course. It is certainly possible. The only difference of opinion I think that Secretary Klinger and I would have is the likelihood of success. He would see the glass as half empty; I would see it as half full.

The point made at the outset by the Chairman, however, I think is one that also has to be kept in mind. Very advanced Scud capability, the type the North Koreans have evolved and have reasonably close to deployment at the present time, the Taepo Dong 2, is generally given credit for capability of reaching only into the mid-Pacific. If the assumed warhead was not in the thousand pound class but substantially smaller, that system could potentially loft packets into orbit; and, in particular, it could throw a small nuclear warhead over the United States.

Such small nuclear warheads exist; there can be no denying. Whether or not the North Koreans have access, by purchase or whatever other means, to the technology that tells them how to make those warheads is any person's guess. But the Chairman's suggested scenario is certainly not conjectural. It has a very substantial character in capability that might exist at the present time.

Mr. EHLERS. Reclaiming my time, just one final question, certainly no scientific component to this one.

But the question is, why would any nation choose to launch a strike of this sort as a single shot? It seems to me that one might do it as a first shot out of a multiple barrage where you are trying to disable the enemy beforehand, but it seems most of the concern we have currently is about rogue nations launching a few nuclear missiles at our country or perhaps at our troops elsewhere.

The question is, why would they launch EMP attacks rather than launching a direct attack against some of our defense installations or our cities?

Dr. WOOD. As I said in my opening statement——

Mr. EHLERS. That question goes to anyone.

Dr. WOOD. As I said in my opening statement, the Soviets planned a very extensive EMP laydown over the United States and elsewhere; and they had a substantial amount of their ICBM force devoted to doing that. So, yes, indeed, they would do just as you anticipated. The rationale thing is punch and punch again and keep on punching so there is a very high likelihood of nothing escaping.

The only reason that you would not do that if you were a Third World power is you lacked both the warheads and the delivery systems that the Soviet Union enjoyed. If you had a few or perhaps only one or two nuclear weapons, you probably would want to use them in the fashion which imposes the largest damage expectancy on the United States and its military forces.

If you are going to go after the military forces and you only have a few, by far and away the most effective way that you could potentially use it is an EMP laydown. If you were going against the American civilization itself, again, the largest damage you could expect to see by far is that associated with EMP laydown.

As I said earlier, a large laydown over the lower 48 States has a damage expectancy which can be reckoned in trillions of dollars. Not 10 trillion, but well above a trillion dollars. So what you get the most bang for your nuclear buck out of, you get it out of most heavily damaging your adversary in either the military sense or the sense of civilian infrastructure. EMP is the attack mode of choice.

Mr. EHLERS. The question still remains. Why would someone do this against the strongest nation in the world with the most missiles available to respond? We used the philosophy of mutually shared deterrence for a number of years. Why would any nation want to tackle us, knowing they would likely be obliterated as a nation?

Dr. WOOD. The Soviet Union, of course, was led by cautious folks. But they didn't do such a thing.

We worry nowadays about what the former chairman of the parent committee, Les Aspin, referred to as the undeterables in the world, the folks who, like Muammar Qadhafi of Libya said, "if I had ICBM's, they would be unleashed against New York and Washington"; against the Kim Chong-ils of North Korea, whose grip on reality is not considered to be an extremely strong one and who apparently continues to contemplate what would have the character of a suicide attack against South Korean and American forces. We worry about people who don't calculate or look at the world in anything like the fashion that we could and simply might not be deterable.

Mr. EHLERS. Thank you very much, Mr. Chairman.

Mr. WELDON. Thank you, Mr. Ehlers, for being with us.

We are going to get to the classified portion; but before I do, I will ask Mr. Pickett if he wants to make any comments.

I would like to make two statements. One is a result of, as the members know, I try to avoid extremes on the issues we deal with, which is very easy, because we have people on both sides, one side that wants to recreate the cold war and the other side that wants to deny there is any threat. I try to find the center in each of the areas we focus on as a subcommittee and full committee.

A couple of comments came to me. One is the status about the control of nuclear weapons. I forget who made the comment in their statement, but I would just mention this.

Doing a lot of work in Russia—I was in Moscow 6 weeks ago with five other Members of Congress in a meeting with General Lebed. General Lebed was asked about the status of nuclear forces in Russia. He told us—using his exact quote, "Mr. Congressman, when I chaired the Defense Council for Boris Yeltsin, one of my responsibilities was to account for all of the suitcase-sized nuclear devices that we built inside of Russia."

He went on to say that during the tenure of building these devices, they had, in fact, manufactured 132; and he told us they had only been able to account for 48.

The status of the others? Who knows. But the point is that Russia certainly doesn't have total control over its nuclear weapons or its fissile material. We have seen evidence of that from time to time. It doesn't mean we should abandon them, but we have to understand and deal with that potential threat to our security and our troops.

The second point is a specific response to the testimony of General Marsh again on page 6 and 7 where you talk about the fact that this likelihood of an event with EMP is unlikely and difficult to achieve, so it doesn't warrant serious discussion. You go on to say the administration's policy is to prevent proliferation and unauthorized access.

My problem with that is—and I am going to ask unanimous consent to enter this as part of the record.

[The information referred to can be found in the appendix on page 109.]

Mr. WELDON. A statement before a congressional committee by retired Lt. Gen. Robert Schwitzer, U.S. Army retired, who as I understand has now been commissioned by Secretary Cohen to do the first white paper on EMP—so I would assume he has some credibility. Secretary Cohen has asked this individual to do a white paper on EMP.

In your statement, General, you make the point that proliferation is, in fact, the way we deal with this.

But if I read, as I am going to do, and enter in the record, as I have already done, the statement by General Schwitzer, it doesn't coincide with that. In fact, General Schwitzer goes into a statement saying, from unclassified sources—quoting him now—"we know that Russia, Ukraine, United Kingdom, China, Australia, France are all ahead in the field of radio frequency weapons, while Germany, Sweden, South Korea, Taiwan, Israel are emerging and have details of the Russian work and proceedings of more than 20 years of international conferences."

He goes on in his statement, which is going to be in the record, to document conferences that have been held where this technology has been distributed and goes on to point out, at some conferences, both Iran and Iraq have been participants and, in fact, have had this technology made available to them.

But here is the rub: Now here you have a supposed expert on this issue developing a white paper for Secretary Cohen who makes two points that our technology control is nonexistent. Yet your point is that is the way the administration deals with this issue, through technology control and control of proliferation.

In fact, let me quote again from General Schwitzer: "Although RF weapon components are on the critical technologies weapons list, there are no up-to-date DOD guidelines or directives on this subject."

This is not me saying this. This is not Lowell Wood. This is the person that the Secretary of Defense has gone to to advise him on this issue. He is saying, there are no up-to-date DOD guidelines—quoting him here—on directives on this subject. An attempt to do so was made 2 years ago when little was known on the subject. As a consequence, decisions within the U.S. scientific community are becoming harder and dicier to make. There is a lack of clear policy guidance and direction.

That is my concern. Any response?

General MARSH. Mr. Chairman, I have reviewed General Schwitzer's testimony carefully; and I think we are talking two different subjects. His entire concern has to do with RF—conventionally generated RF weapons—that is, conventionally explosive generated RF weapons; and he discusses that at great lengths. I don't believe—I don't think nuclear EMP and that are two different subjects.

Mr. WELDON. Thank you.

I would say, on page 5, you say, we have also looked at localized RF weapons and discussed them with RF weapons experts. It is theoretically possible to develop such weapons; but, to my knowledge, the practicality has not been demonstrated. We will get into that in the classified session.

Maybe I am misreading this, but the general thrust of your statement is the administration's policy is to prevent proliferation and unauthorized access, which I assume you meant to include both RF weapons which you did discuss in the preceding paragraph as well as EMP, is that correct?

General MARSH. That is not correct. I did not intend that to cover RF weapon technology.

Mr. WELDON. Do you agree with General Schwitzer we do not at this time have clear policy directives within DOD on RF weapons technology?

General MARSH. Sir, I don't know. I did not address whether there is a DOD policy on RF weapon technology or not. I just can't answer that.

Mr. WELDON. Does anyone know, in terms of technology transfer?

Mr. KLINGER. Sir, the only thing I can address is, as General Schwitzer correctly pointed out, those things are controlled. It is a critical technology.

We have the policy guidance that exists with regard to space systems which is not explicit with regard to RF weapons. It is generalized with regard to the classes of threats against which we need to protect ourselves which encompass RF weapons.

Those are the only two data points I have.

Mr. WELDON. He goes on to give a very specific example that I am not going to get into about Weltron microwave tubes and an example of that not being properly controlled by DOD, and that is my concern.

Again, not to scare anyone, but from a practical standpoint, I agree we should be controlling proliferation. But if we have someone, also an expert, saying we are not doing that, OK, if it is only RF weapons, then we need to discuss that.

That problem exists with EMP technology—in his case, he is talking about RF and, to some extent, the impact of a laydown. That is the point of what we are trying to get at. From a policy standpoint, are we doing enough? Do we need to do more? Is there adequate testing in place? Perhaps most importantly, which will be the subject of the next part of this hearing, is the intelligence community adequately updating us with an assessment that we can make an intelligent decision upon where to base our dollars?

Our initial—my initial feeling on that is when we contacted them and got a 10-year-old report that perhaps maybe that is not the case. Maybe we will hear differently when they come in in the next session. But that is going to be the subject of the third part of this hearing, which will take place as soon as I conclude this.

Mr. Pickett, any closing comments?

Mr. PICKETT. One issue, Mr. Chairman, that is not clear to me; and that is whether, in assessing the issue of the nuclear threat, that this notion of using a nuclear device to create EMP is classified the same as using a nuclear device for any other purpose against our Nation.

I don't know if anyone is prepared to answer that this morning. But it seems to be, inferentially anyway, if you start messing around with nuclear devices, whether it is an EMP generator or whether it is actually something that is intended to destroy physically property on the face of the continental United States, that the response is going to be the same.

Does anyone have any information they can give us on this issue?

Dr. WOOD. Yes, sir, but I would respectfully suggest it is appropriate for closed session.

General MARSH. I would only add one comment, and that is I think the uncertainty of the results of a deliberate EMP attack, especially by a terrorist or some rogue nation, the uncertainty of what they would achieve I would think would mitigate strongly against using that one or few weapons for its EMP effect versus known effects of very serious damage by way of blast and thermal. That is a personal opinion.

Mr. PICKETT. Thank you very much.

Mr. WELDON. Let me thank all of our witnesses for outstanding testimony and helping us better understand what is arguably a very complicated subject. With the exception of our very technical people that you have seen today, most of us really are not adequately capable of assessing the technical aspects but rather, from a policy standpoint, want to make sure we are taking every step we can to deal with this emerging threat. We appreciate your help in accomplishing that.

You are all invited to stay because you have the appropriate clearances, I am sure, as we adjourn this session of the R&D subcommittee.

The subcommittee stands adjourned.

Staff will clear the room so we can have a classified briefing with the individuals that I have outlined: Dave Osias, the National Intelligence Officer for Strategic Systems and Nuclear Proliferation; Dr. Jose Pina from the Central Intelligence Agency; and Dr. Nelson DeGangi from the Defense Intelligence Agency. That classified session will be in room 2212.

[Whereupon, at 12:47 p.m., the subcommittee proceeded in Executive Session.]

APPENDIX

JULY 16, 1997

Opening Statement
Rep. Curt Weldon
Chairman, Military Research and Development Subcommittee
Hearing on the Effects of Electromagnetic Pulse
July 16, 1997

The subcommittee will come to order.

This morning, the Military Research and Development Subcommittee meets to receive testimony on the effects of electromagnetic pulse, or EMP, on our military systems and civilian infrastructure. I have also invited the members of the Procurement Subcommittee to attend the hearing, many of whom have expressed an interest in this subject. Welcome to my colleague and good friends Duncan Hunter, chairman of the Procurement Subcommittee and ranking member Ike Skelton, and to all my friends from the Procurement Subcommittee.

[We are also honored to have Mr. Floyd Spence, Chairman of the full committee, with us today, and our good friend and ranking member, Mr. Ron Dellums.]

Committee members were invited to a June 26 briefing by representatives of Applied Physics Lab on the impact of an electromagnetic pulse generated by a high altitude nuclear blast. The Subcommittee is meeting here today to explore in more detail EMP effects on our military systems and the civilian infrastructure, how confident we are that we can predict these effects, our potential vulnerabilities, what policies and practices guide our efforts to protect our systems, and the steps we have taken and can take to ameliorate these vulnerabilities.

EMP can be generated in several ways, but the widest effects are caused by a high altitude nuclear blast. All of us here understand that the threats posed to our military systems and civilian infrastructure by high altitude EMP are not new. Atmospheric nuclear tests in the 1950s and 1960s revealed a number of then-unanticipated results, including electrical and communications disruptions hundreds of miles from the test sites.

There may be, however, new dimensions and vulnerabilities that we need to look at more carefully. In the 1950s, electronic systems used vacuum tubes, not very sophisticated, but resistant to EMP. Today, computers with more and more microcircuits packed into smaller and smaller chips are key to the efficiency of virtually all commerce in the U.S. These have much more computing power, but are also much more sensitive to disruption and more easily disabled by EMP. As our reliance grows, so does our potential vulnerability.

Likewise, potential military vulnerability may be growing. The revolution in military affairs has brought with it a much greater dependence on information technologies. The ability to generate raw data, process it into useable form, and communicate information to the right people and systems is critical to military success. Yet the sensors, computers, and communications assets essential to this revolution could be vulnerable. For example, ninety five percent of our military communications go through commercial channels. Are we confident that EMP won't disable or disrupt these commercial communications systems? How confident are we that the military could continue to communicate effectively if commercial systems are disrupted or disabled by EMP? How thoroughly do we protect our weapons systems from EMP? Are we confident they will continue to function?

At the same time our vulnerabilities may be increasing, the subcommittee is worried about complacence in this area. It is true that the Cold War is over, and the threat of a deliberate nuclear attack by Russia is very low. That does not necessarily mean we have nothing to worry about. The proliferation of nuclear weapons and ballistic missiles continues. If we come into conflict with a rogue nation, in possession of just a very few nuclear weapons, our adversary may use, or threaten to use, one of those weapons to blind the US military or to damage our economy or that of an ally. Yet when we invited the intelligence community to provide us with their most up to date material on EMP threats, we received documents that were ten years old. We understand that no NIE has been issued on EMP since the 1980s. We also understand that some of our EMP testing assets have been dismantled or are in disrepair.

I'd like to welcome our witnesses today and thank them for taking the time and effort to help us understand these issues. To talk about EMP effects, we have Dr. Gary Smith, the Director of the Applied Physics Lab from Johns Hopkins University, and Dr. George Ullrich, the Deputy Director of the Defense Special Weapons Agency. To talk about the impact of these effects on our military systems and commercial infrastructure, we have General Robert T. Marsh, US Air Force retired, chairman of the President's Commission on Critical Infrastructure Protection; Hon. Gilbert I. Klinger, Acting Deputy Under Secretary of Defense (Space); and Dr. Lowell Wood of Lawrence Livermore Laboratory. From the intelligence community we have Dr. David Osias, the National Intelligence Officer for Strategic Systems and Nuclear Proliferation, Dr. Jose Pina from the Central

Intelligence Agency, and Dr. Nelson Degangi from the Defense Intelligence Agency. Gentlemen welcome, and thank you again.

Dr. Smith, please proceed.

TESTIMONY

DR. GARY L. SMITH
DIRECTOR

THE JOHNS HOPKINS UNIVERSITY
APPLIED PHYSICS LABORATORY

SUBCOMMITTEE ON MILITARY RESEARCH AND DEVELOPMENT
HEARING ON ELECTROMAGNETIC PULSE

HOUSE NATIONAL SECURITY COMMITTEE

July 16, 1997

Effects from Nuclear Explosions at High Altitudes

Testimony by Dr. Gary Smith to the House National Security Subcommittee on Military Research and Development on 16 July 1997

Introduction

Good Morning,

Thank you for the opportunity to testify before the Subcommittee on Military Research and Development. I am Gary Smith, the Director of The Johns Hopkins University Applied Physics Laboratory, located in Howard County, Maryland.

The Applied Physics Laboratory operates under about a dozen task order contracts with a number of major sponsors, covering between two and three hundred separate tasks in any given year. All of our funding is derived from programs. About 80% of our funding comes from sponsors within the Department of Defense and about 15% from the National Aeronautics and Space Administration. The rest comes from various government departments and agencies.

Recently Congressman Bartlett asked me to advise him on the subject of this particular hearing, the electromagnetic pulse produced by a high altitude, nuclear explosion, and the implications for defense systems and capabilities, and civilian infrastructure. My staff has completed an assessment and I am prepared today to present the results.

In short, we have found that the phenomenon is very real and well understood by the nuclear weapons effects community; that our strategic systems, and their command, control and communications infrastructure, have been designed and built to survive and operate effectively in such an environment; that there would likely be pronounced effects on the civilian infrastructure from such a pulse; that the magnitude and extent of these effects is difficult even to estimate; and that it is probably not feasible to completely protect the entire infrastructure from the effects of such a pulse.

Discussion of the effects of a nuclear attack on the United States has, in the past, tended to focus on the effects such attacks would have on our defense capabilities and on our ability to function militarily in such a situation and to prevail on favorable terms. Mutual understanding by the United States and its adversaries of the widespread and cataclysmic damage that hundreds of megatons of nuclear explosions would cause to our society and its infrastructure, and the corresponding damage a like response would cause an adversary, was the centerpiece of the philosophy of the Cold War: Mutual Assured Destruction.

The Cold War is over, but the threat from nuclear explosions remains. This could result from a coordinated attack using substantial numbers of modern weapons or from an attack involving, at most, a few crude weapons in the hands of terrorists or rogue nations. The latter scenario is of particular concern in this post-Cold War era. In this testimony, I will discuss the effects of distant nuclear explosions on electronic systems and, more specifically, the implications of these effects on the functioning of our civil infrastructure--one that is today almost totally dependent on increasingly sophisticated and perhaps also increasingly fragile electronic systems and subsystems.

The immense amount of energy liberated by a nuclear explosion, principally in the form of X-rays, gamma rays, and high-energy neutrons, can produce a wide range of effects on this infrastructure. The well-known effects of a near-surface or ground burst (which include blast, ground shock, and thermal radiation) are actually indirect: they result from conversion of the bomb's energy into thermal and kinetic forms. Effects of a nuclear detonation at high altitude on satellites and missiles in flight, however, are direct: the energy from the detonation interacts in its original form with the target system to induce malfunction or damage. An additional, and very important, effect of a high-altitude detonation, particularly for airborne and ground systems, is the high altitude electromagnetic pulse (HEMP) that results from the conversion in the earth's ionosphere of weapon gamma-ray energy to radio frequency energy that propagates toward the earth's surface.

To understand the potential threat to unprotected* electrical and electronic systems better, it is useful to limit the present discussion to those problems that occur as a result of nuclear explosions in and above the atmosphere--often referred to as *high-altitude* nuclear explosions, with the burst altitude generally measured in the hundreds of kilometers--although some very devastating, albeit relatively local, electrical effects can occur as a result of nuclear explosions at or near the surface of the earth. It is also useful to discuss HEMP and satellite disturbances separately and to limit the discussion to the effects of a single explosion. Multiple explosions considerably complicate the environment and can make the resultant effects more severe. No attempt is made here to describe the additive effects of multiple explosions.

In this testimony, I will describe the commercial infrastructure as an interconnected communications network made up of ground systems (e.g., computers controlling a financial network or the electronic switches that operate a telephone system) and the satellites that support radio communications among the ground systems. I will first consider electromagnetic pulse (EMP) phenomena, system vulnerability, and system protection and identify specific EMP-related issues for ground system components of the civilian infrastructure. I will then discuss nuclear threats to space-based elements of the infrastructure, specifically reviewing threat environments and the effects of prompt and delayed radiation exposure on satellite systems.

---

*By "unprotected", we mean that no special effort beyond good engineering practices has been made to shield electronic equipment from the effects of nuclear explosions.

Electromagnetic Pulse (EMP)*

The detonation of a nuclear weapon produces high-energy gamma radiation that travels radially away from the burst center. When the detonation occurs at high altitudes where the mean free path of the gamma photons is very long, these photons travel long distances before they interact with other particles. Gamma rays directed toward the earth encounter the atmosphere where they interact with air molecules to produce positive ions and recoil electrons which are called Compton electrons after the man who discovered the effect. The Compton recoil electrons also travel away from the detonation point but are deflected by the earth's magnetic field. Figure 1 depicts this situation.

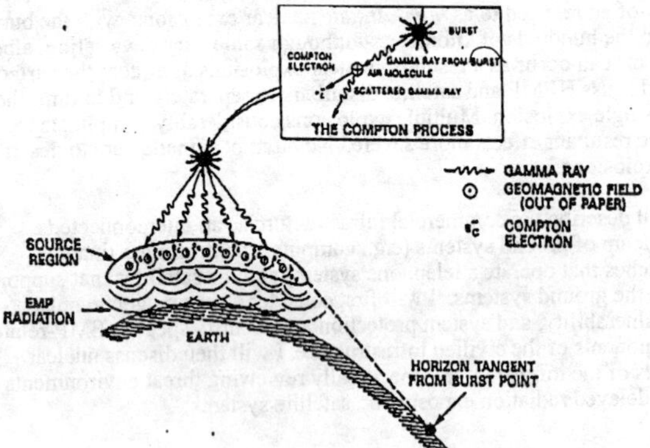

Figure 1. Mechanisms of High Altitude Emp (HEMP) Generation

---

*Much of the discussion regarding the nuclear electromagnetic threat and its effects was taken from the article "Understanding the EMP Threat" by G. K. Soper and K. F. Casey that appeared in the November 1987 issue of *Defense Electronics*.

The gamma radiation interacting with the air molecules produces charge separation as the Compton recoil electrons are ejected and leave behind the more massive, positive ions. The earth's magnetic field interaction with the Compton recoil electrons causes charge acceleration, which further radiates an electromagnetic field. HEMP is produced by these charge separation and charge acceleration phenomena, which occur in the atmosphere in a layer about 20 kilometers (km) thick and 30 km above the earth's surface. The effective source region covers the earth within the solid angle subtended by rays from the detonation point that are tangent to the atmosphere. The fraction of the Continental United States affected by EMP is dependent on the height of the burst as depicted in Figure 2. To an observer on the ground, the incoming wave appears to be a plane wave propagating toward him from above the detonation point. The amplitude, duration, and polarization of the wave depend upon the location of the burst, the type of weapon, the yield, and the relative position of the observer. The electric field resulting from a high-altitude nuclear detonation is on the order of 50 kilovolts per meter (kV/m), with a rise time on the order of 10 nanoseconds (ns) and a decay time to half-maximum of about 200 ns (Figure 3). By comparison, a localized lightning strike 10 meters away has a higher peak amplitude, but it rises more slowly than the EMP peak. Therefore, protection may be available.

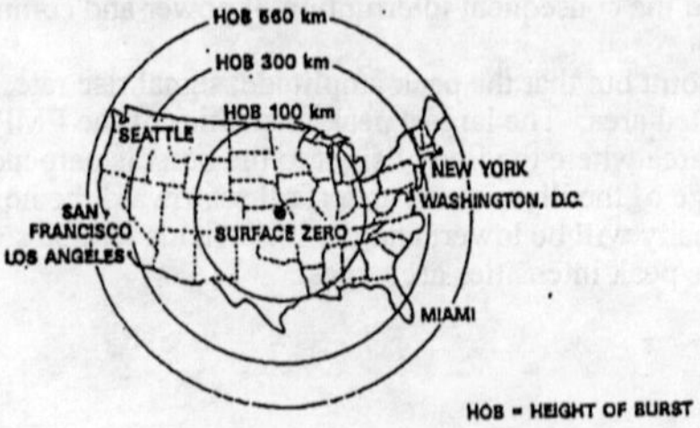

Figure 2. Extent Of HEMP Coverage

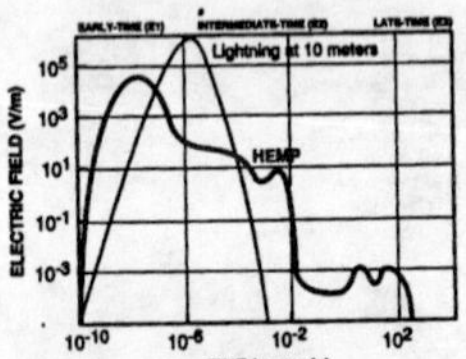

Figure 3. HEMP Time Regimes

HEMP consists of three distinct time or frequency regimes. The acceleration or rotation of the electrons around the magnetic lines of force produces the fast-rise pulse termed early-time pulse, or E1.

The intermediate time, or E2 portion, of the HEMP waveform is generated by electrons produced by delayed gamma rays. Some gammas are delayed by scattering collisions, and some are generated by inelastic neutron interactions with air molecules. The E2 portion of the HEMP is much less intense than E1.

Expanding weapon debris forms a highly conducting plasma sphere that interacts with the earth's magnetic field. Since magnetic lines of force cannot pass through this conductive region, they are forced to expand outward with the weapon debris creating a very low-frequency, low-peak value EMP termed variously the magnetohydrodynamic (MHD) signal, the late-time pulse, or E3.

Each of these three time regimes is important because each contains different frequencies, and HEMP energy will couple most efficiently with conductors matched to these frequencies. E1 creates the most intense fields and couples well to local antennas, equipment in buildings (through apertures), and conducting lines of any length. E2 couples well to overhead and buried conductive lines about 1 km long. E3 couples well to conductors longer than 10 km, such as power and communications networks. E3 is capable of generating potential differences of a few kilovolts between the ends of these lines; the resulting upset of the lines' voltage balance can lead to their shutdown and the consequent interruption of power and communication services.

It is important to point out that the peak amplitude, signal rise rate, and duration are not uniform over the illuminated area. The largest peak intensities of the EMP signal occur in that region of the illuminated area where the line of sight to the burst is perpendicular to the earth's magnetic field. At the edge of the illuminated area (farthest toward the horizon as seen from the burst), the peak field intensity will be lower; and the EMP fields will be somewhat longer-lasting than in the areas where the peak intensities are largest.

System Susceptibility To EMP

An electromagnetic field interacts with metallic conductors by inducing currents to flow through them. A television antenna, for example, is a collection of metal conductors arranged to facilitate the induced current flow in the frequency range allocated for television broadcasting and to transfer the signal to the receiver. Other conducting structures such as aircraft, ships, automobiles, railroad tracks, power lines, and communication lines connected to ground facilities also effectively serve as receiving antennas for EMP coupling. If the resulting induced currents and voltages - which can be large - are allowed to interact with sensitive electronic circuits and components, they can induce upset in digital logic circuits or cause damage to the components themselves.

Ground facilities, for example, those housing the large computers central to the functioning of our financial system, are typically nodes in a larger network and are connected to overhead or buried cables for power and communication. They are also connected to buried pipes for water supply and waste disposal and are typically equipped with communication antennas and distributed security systems of various types. All of these features can direct EMP energy into the facility. Analyses and simulated EMP testing have shown that currents carried to a facility by long overhead or buried conductors can reach thousands of amperes. Shorter penetrating conductors can carry hundreds of amperes into a facility. Direct EMP penetration through the walls and windows of an unshielded building can induce currents of tens of amperes on illuminated interior conductors.

When EMP energy enters the interior of a potentially vulnerable system, it can cause a variety of adverse effects. These effects include transient, resettable, or permanent upset of digital logic circuits and performance degradation or burnout of electronic components. The collected EMP energy itself can cause malfunction or device failure directly; or it can trigger the system's internal power sources in unintended ways, causing damage by the power sources within the system itself.

Unprotected systems attached to long lines, such as communication and power systems, experience HEMP-induced voltages and currents whose intensities vary with the length of the lines. Intensities can be severe (approximately 1.07 megavolts (MV) or 2.7 kiloamperes (kA) for a 20-kV/m electric field). The chart shown in Figure 4 can be used to determine the maximum open-circuit voltage and the short-circuit current induced on an overhead power line or cable on the ground.

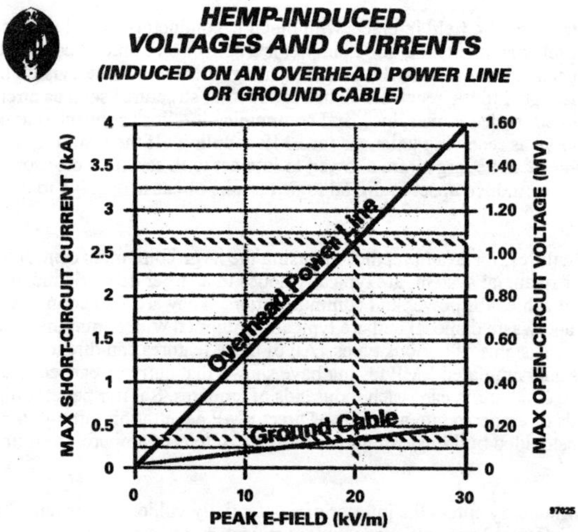

Note: This chart is to be used for conceptual purposes only.

Figure 4.

Sample problem: A high altitude burst produces a 20-kV/m E-field that couples to an overhead power line and a cable on the ground. Find the maximum open-circuit voltage and short-circuit current for both the power line and ground cable.

Answer: The open-circuit voltage and short-circuit current are found by plotting a perpendicular line from a point at 20 kV/m on the x axis (peak E-field) to the intersecting points for both curves. The open-circuit voltage is read on the right side of the chart (approximately 1.07 MV for a power line and 0.14 MV for a ground cable), and the short-circuit current is read to the left of the chart (approximately 2.7 kA for a power line and 0.3 kA for a ground cable).

In summary, EMP introduces two collectively unique features to the overall picture of system susceptibility to nuclear effects. These features, taken together, distinguish EMP from all other forms, both natural and manmade, of electrical stress and response: First, stresses induced by EMP can significantly exceed those ordinarily encountered in system circuits and components and can thereby increase the probability of upset and burnout occurring in electrical and electronic systems. Second, EMP can cause this increase to occur nearly simultaneously over a large area: about one million square kilometers for a high altitude burst. These unique features, together with the lack of occurrence of EMP-like phenomena in the normal day-to-day environment, cause great difficulty in attempting to deal with EMP as a normal engineering problem. In particular, EMP can induce *multiple, simultaneous* upsets and failures over this wide area. It should be noted, however, that while these unique characteristics do not make unacceptable electrical or electronic malfunctions inevitable, neither is the avoidance of such malfunctions automatic.

EMP Vulnerability

Electronic systems have tended to become more susceptible to EMP over the years, largely as a result of the advances in electronics technology made since the development of the transistor. The current and voltage levels associated with the normal operation of electronic devices and the power or energy levels at which failure can occur have all fallen steadily as solid-state and integrated-circuit technologies have placed ever-increasing numbers of devices and circuits on semiconductor chips. In addition, the increasing use of digital circuitry to perform more complex functions has added to the risk and the seriousness of the consequences of digital logic upset. As anyone who has experienced a momentary power fluctuation while using a personal computer will attest, such upsets can readily occur with unpredictable, and usually unfortunate, results.

There is a large body of evidence to suggest that unprotected systems can be vulnerable to EMP-induced effects. The primary evidence for susceptibility of modern electronics systems to EMP has been obtained from tests where representative systems have been exposed to simulated EMP environments or EMP-induced stresses. This kind of testing is typically performed using EMP environment simulators, where an EMP-like electromagnetic field is generated with conventional pulse-power sources and used to illuminate the test object, or testing is performed using direct drive of antenna leads.

EMP simulation testing has shown that unprotected systems frequently experience both permanent damage and transient upset when subjected to EMP-like stresses. Temporary system outages, circuit upsets, and permanent failures of semiconductor devices have all been observed in tests of unhardened military equipment and command, control and communications (C3) systems. In view of our inability to predict the occurrence of damage or upset in systems that were not specifically designed to be "hard" to EMP, reliable conclusions concerning the EMP survivability of a specific system cannot be drawn in the absence of a detailed test and evaluation of that system. The general pattern, however, is clear. Protection from EMP effects is necessary for critical systems, and high confidence in system hardness can be gained only through testing under conditions that closely approximate the threat.

Electromagnetic Pulse Protection

Protection or hardening of potentially vulnerable system elements from the effects of EMP requires that these elements be shielded from the EMP field and from potentially dangerous EMP-induced currents and voltages. To put the problem into perspective, consider that the high-altitude EMP can induce open-circuit voltages on the order of megavolts or short-circuit currents on the order of 10 kiloamperes on overhead conductors such as power lines. Small-signal electronic circuits typically operate at levels of a few volts or a few tens of milliamperes. To prevent EMP-induced transients on power lines from producing upset in these circuits, sufficient protection must be provided to reduce by at least seven orders of magnitude the transient peaks induced on the power lines to levels that can be tolerated by the small-signal circuits operated from these lines. Certain sensitive circuits, such as magnetic disk read-head amplifiers, require even more isolation.

All approaches to the protection problem are based on the idea that a barrier must be erected between the stresses induced by the EMP and the system elements that these stresses can adversely affect. The various approaches differ principally in the choice of barrier location.

A completely closed (integral) and perfectly conducting shell around a region of space totally excludes externally generated electromagnetic fields from the enclosed region. This is the basis for the integral-shielding approach to EMP protection: One effectively wraps the system to be protected in a metallic shield without apertures or other penetrations (a "Faraday cage") and thereby isolates it from the hostile electromagnetic environment. For any system to be useful, however, it must interact with the outside world. A pilot must be able to see through cockpit windows, communications must be maintained, and manned ground facilities must be supplied with air and water and often with externally generated power. For these reasons, the integral shielding approach to EMP protection is modified to permit, and to protect, those penetrations that are necessary to the system's function while closing all unnecessary penetrations. The means employed for effective penetration protection include filters and surge arrestors for line penetrations such as those associated with antenna and power connections; wire-mesh or transparent conductive-film coatings for windows where visibility must be preserved; metal honeycomb for ventilation ports; and conducting gaskets for doors and hatches.

This "integral shielding with penetration control" approach to system protection has proven most attractive for hardening many classes of military systems and C3 facilities. A new strategic missile system, for example, is an ideal candidate for this form of EMP protection. The fact that the system is *new* means that EMP protection can be incorporated into the entire system acquisition process from concept definition through production and deployment. The fact that the system is *strategic* implies that it is a high-value asset; and as a consequence, a low-risk, high-confidence hardening approach is necessary. Furthermore, long-term hardness maintenance is required. Finally, the fact that the system is a *missile* implies that the structure is relatively compact and enclosed, possessing few necessary penetrations that require protection. Because the number of necessary protection measures is small, system hardness will be relatively easy to validate and maintain over its life cycle. This is definitely not the case, however, for a system of continental size such as a commercial banking network, a transportation system, or an interconnected telephone network.

At the other end of the spectrum of EMP protection lies hardening at the individual-element level. In this approach, the barriers between the EMP stresses and the potentially vulnerable elements are chosen close to the elements themselves. Local rather than global shielding is used; and individually hardened piece parts (diodes, transistors, and integrated circuits), hardened circuit and fault-tolerant logic design, and terminal protection devices (filters and limiters) are employed to protect the system elements. The number of such elements, however, can be extremely large; and the number of protection measures that must be applied to the system will be correspondingly large. As a consequence, hardness validation and maintenance can be difficult when individual element hardening is used. Furthermore, the risk of incomplete system protection can be substantial.

Even given these shortcomings, individual-element hardening is often the most expeditious approach for adding EMP protection to an existing facility or complex. The risk of incompleteness of the protection will be higher than for an integral shield with protected penetrations; and the increase in system hardness may not be quantifiable; but the cost will be relatively low. If the most important penetrations and the most critical electronic boxes can be identified and protected--and engineering judgment and experience will be crucial in making the necessary identifications--one can reasonably expect to achieve a significant reduction in the vulnerability of the system. As the system becomes more complex, however, the confidence in this approach degrades rapidly.

Some systems such as civil telecommunications networks would be very difficult and expensive to protect with high confidence because of their great spatial extent and the diversity of their elements. Because of its obvious importance, the robustness of the U.S. telecommunications system to the EMP threat has been the object of a great deal of attention. Beginning in the late 1960s, the Department of Defense has sponsored several assessments of the EMP hardness of leased portions of this network. No conclusive evidence has been found that the system would--or would not--fail catastrophically under EMP-induced stress, and there is no way that the system can be realistically tested. There, we cannot be confident that the network will function reliably after exposure to one or more high-altitude nuclear bursts.

Threats to Space-Based Elements

Satellites play an essential role in U.S. military strategic and tactical communications by enabling critical command, control, detection, early warning, and precise timing functions. Satellites have also become key elements in our civil telecommunications industry. Space has become crowded, and slots for geosynchronous satellites are at a premium. Interconnected low-orbit systems made up of multiple satellite nodes will soon give the user the capability to instantaneously connect to other users worldwide. Parts of this space-based infrastructure could be put at significant risk in the event of one or more nuclear detonations in space.

Satellites are not inherently hard, and the hardening levels that are economically and technically feasible are not sufficient to protect them against a determined direct attack. Furthermore, depending on certain factors, these hardening levels may not be sufficient to prevent disruption as a result of "accidental" exposure to nuclear radiation from explosions in space whose primary purpose is to create other effects such as EMP on the ground. Many dedicated military satellite command, control, communications, and intelligence systems include the requirement for nuclear hardness. Although hardening alone does not necessarily result in increased survivability of satellites, satellite hardening can reduce the vulnerability to high-altitude nuclear detonations associated with a terrorist or rogue-state attack or with a ballistic missile defense exchange. Satellite hardening can also prevent the elimination of many satellites by the detonation of one or a few well-placed weapons in space. The value of these satellites, in terms of both replacement cost and the cost of disruption of a major part of our information age infrastructure, is impossible to calculate.

For commercial systems, one can reasonably expect a degree of protection from distant explosions (at distances of some thousands of kilometers) as a result of the application of the standard engineering design and construction practices that are used in modern satellite systems. Commercial systems must, after all, operate reliably for long periods of time in space where even the natural environment can cause performance problems as a result of accumulated radiation. Satellites must cope as well with the unexpected stresses associated with high-energy radiation and electromagnetic energy flux from intense solar storms. Without a dedicated analysis and testing program, however, it would be difficult to confidently quantify the level of inherent protection that can be achieved in commercial systems. Until such a quantification can be made, it is difficult to predict adequately how well these systems will perform when subjected to a nuclear environment.

Threat Environments

Satellites can only be affected by detonations at high altitudes. For low-altitude bursts, the atmosphere itself greatly shields the satellite from the nuclear radiation. In general, the detonation altitudes at which satellites can be affected are not well defined because they depend on the attenuation of the atmosphere, the range of possible weapon yields, geometrical factors, and satellite sensitivities. A reasonable rule of thumb, however, is that for nuclear detonations below about 40 km, satellites, particularly communications satellites, will not be affected by direct radiation. The high-altitude regime of interest therefore excludes ground and air bursts. Included in this altitude regime are the increasingly possible high-altitude detonations that could occur as a result of a determined terrorist or rogue-state attack.

A variety of stresses can be induced on satellite systems by high-altitude nuclear explosions. A satellite in the line of sight of a nuclear detonation can be exposed to various types of nuclear radiation, including neutrons, gamma rays, and X-rays. This prompt radiation leads to effects that become more significant as the distance between the weapon and the satellite is reduced. Electrons produced by the radioactive decay of fission debris can be captured by the earth's magnetic field and can repeatedly bombard any orbiting satellite traversing the artificially enhanced magnetosphere. This delayed radiation produces high doses in electronics that can result in permanent damage to critical components, reducing the satellite's functionally useful lifetime.

In the subsequent discussion, nuclear effects on satellites are categorized as either "prompt" or "delayed". The prompt effects occur immediately, result in either upset or permanent damage of electronics, and are caused by direct neutron, X-ray, and gamma-ray interactions with the satellite. The delayed effects occur over a longer period of time, result in a gradual deterioration of the performance of the electronics, and are primarily caused by the enhanced geomagnetically trapped electron environment.

Prompt Radiation Exposure

Figure 5 shows the prompt nuclear environments as a function of satellite-to-burst separation distances. As noted earlier, upset effects are momentary malfunctions of electronic components that cause no permanent damage to the component. Although these perturbations can last much longer than the source radiation that causes them, they presumably could be corrected by ground control if sufficient diagnostic times were available. Permanent damage effects are those that cannot be corrected; these include significant degradation of device operation parameters, destruction of one or more devices, or any other malfunction that permanently prevents operation of the space system or one of its subsystems.

Fluence vs. Range to Effects

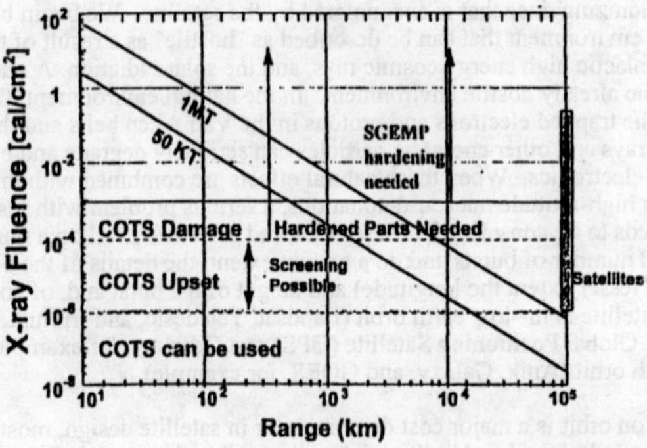

Figure 5.

It can be seen from Figure 5 that X-ray induced shock (so-called "thermomechanical" shock) damage to certain microcircuits, as well as gamma-ray induced ionization upset or burnout, occurs at relatively short ranges. System-generated electromagnetic pulse (SGEMP) refers to the electromagnetic fields generated by the motion of electrons scattered out of (satellite) material by ionizing radiation (primarily X-rays). Threshold levels for SGEMP-induced upset and damage are very difficult to predict. The levels shown are for illustrative purposes only and do not constitute hard, satellite-specific numbers. Instead, the levels shown indicate where upset due to SGEMP fields could occur. One of the most important implications of the data shown in the figure for unhardened satellites is that potentially damaging effects can occur even at radiation levels associated with extremely large burst-to-satellite separations.

For military systems, SGEMP is of particular concern because it is a system dependent effect and one that can occur at very long ranges. The underlying physics is complicated, and SGEMP hardening technology is just now reaching a mature and confident level of application. SGEMP analysis techniques require a high level of computational sophistication and threat-level system tests are required for reliable predictions of system reliability. Even though our ability to test these effects at the system level is severely limited, new concepts have been proposed whereby design principles can be combined with test techniques to maximize system hardness and verify hardness without system-level testing.

Delayed Radiation Exposure

Perhaps a greater threat to our commercial satellite constellations is that from enhanced trapped radiation that results from nuclear bursts at high altitudes. This is not EMP but is an increase in the total ionizing dose that is encountered by the satellite. We begin by discussing the natural radiation environment that can be described as "hostile" as a result of the Van Allen radiation belts, the galactic high energy cosmic rays, and the solar radiation. A nuclear burst simply "enhances" the already hostile environment. In the natural environment, the total ionizing dose from the trapped electrons and protons in the Van Allen belts and the single event effects from cosmic rays and other energetic particles can seriously degrade and even destroy unhardened satellite electronics. When these natural effects are combined with the enhanced radiation levels from high-altitude nuclear detonations, a serious problem with respect to satellite lifetimes needs to be considered. These enhanced effects depend on a number of factors: the yield and number of bursts and, to a certain extent, the details of the bomb design; the latitude (and to a lesser extent the longitude) and height of the burst and, of course, the "kind" of orbit the satellite is in--low earth orbit (Landsat, Teledesic, and Iridium, for example), medium earth orbit (Global Positioning Satellite (GPS) and Odyssey, for example) or geosynchronous earth orbit (Anik, Galaxy, and GOES, for example).

Because time on orbit is a major cost discriminator in satellite design, most U.S. commercial satellites are designed and built with "hardening" to the natural environment as a requirement. Even so, there are examples of satellite failures due to severe magnetic storms. In addition, during the 1962 high-altitude nuclear testing series, a number of low earth orbit satellites failed as a result of the total dose effects resulting from the enhanced trapped radiation. It is not at all obvious that protection to the natural environment will adequately protect commercial satellites from enhanced effects from one or a few high-altitude bursts. The worst scenarios involve high-altitude, low-latitude (~20°) bursts, even at relatively low (~50 kiloton [kT]) yields. Here, low earth orbit satellites (at altitudes less than about 1000 km)-which, in the natural environment, see the lowest natural threat and therefore receive the least hardening-can be subjected to high levels of total dose radiation that can last for months and even years and affect the satellites' useful lifetimes. In the case of high-latitude detonations (>45°), the enhanced radiation is short-lived--perhaps a few days or weeks--and is more or less independent of the height of burst. Although the lifetimes of commercial satellites are highly scenario-dependent, burst(s) at lower latitudes tend to enhance the lower regions of the Van Allen belts and pose a threat mainly to satellites in low earth orbit. For high-latitude burst(s), the higher regions of the Van Allen belts are enhanced, affecting satellites such as GPS satellites and those in geosynchronous orbit. (An excellent article on this subject by R C. Webb, et al., "The Commercial and Military Satellite Survivability Crisis" can be found in *Defense Electronics*, August 1995).

Concluding Remarks

We have described separately the potential effects on electronic systems resulting from a high-altitude nuclear detonation on the two basic elements making up a typical telecommunications-like infrastructure: high-altitude EMP effects on ground systems and; exposure of satellites to, and potential damage from, direct radiation and from delayed effects due to enhanced trapped radiation. As I have stated, the EMP effects occur, for all practical purposes, simultaneously and, depending on the height of burst, can cover the continental United States. This same burst will expose all satellites within the line of sight of the burst to prompt, potentially damaging radiation and will enhance the Van Allen belts that the satellites will continually traverse, giving rise to damaging delayed effects. These wide-area features separate high-altitude nuclear detonations from other weapons of mass destruction in the number of systems that are simultaneously placed at risk.

As was discussed earlier, it is relatively inexpensive to confidently protect electronic equipment at the part, component, subsystem, and even at the system level given that the system is "manageable"—recall the earlier discussion of a strategic missile as a manageable system.* One can also connect "islands" of protection systems together using fiber optics or other protected connections. There is ultimately, however, a limit as to how many of these "islands" can be connected together before the vastness and complexity of the network, and the concomitant difficulty of maintaining an adequate level of protection, make system-wide protection cost-prohibitive or unfeasible from an engineering point of view.

Finally, our opinion is that the coverage and levels that would ensue from an EMP attack are well understood. However, the overall effects on specific terrestrial systems are not as well understood. It is without question that "unprotected" electronic systems must be considered at risk when exposed to the effects of nuclear weapon detonations. Unfortunately, the level of risk and the consequences of continental-wide exposure of our electronic infrastructure are simply not calculable to any degree of certainty. Arguments have been put forth that the complexity, vastness, and redundancy of our electronics infrastructure are not the limiting factors but rather the saving grace. We can be confident that not all systems will fail simultaneously when exposed to a nuclear explosion environment. It is fair, however, to assume that upset and damage will occur. But it is impossible to predict precisely how extensive the damage will be or to predict confidently beforehand whether or not the system will operate adequately after being exposed to this threat.

I hope that I have been able to give you an idea of the phenomenology associated with EMP and the qualitative effects on our civilian infrastructure. I sincerely thank you for the opportunity to address the Committee.

---

* In general, nuclear hardening costs represent a small percentage of total system cost. See, for example, an excellent summary of this topic by J. M. Pierre et al., "The Cost of Radiation Hardened Systems," a paper based on a presentation at the Government Microcircuit Applications Conference, March 1996. In this paper, they suggest that satellite system costs driven by strategic X-ray hardening requirements are on the order of 5% or less of the total cost, and that hardening a satellite to the natural environment typically costs less than about 1% of the total. Recent costs studies have also shown cost deltas for EMP hardening of ground systems to be about 2% or less of a hardened ground-based system's total cost. As they point out, these minimal cost hardening deltas depend upon defining the threat early in the system design; designing radiation hardness into the system from the beginning; and, most importantly, the availability of a domestic radiation-hardening technology infrastructure.

Biography of Dr. Gary L. Smith

Dr. Gary L. Smith became the sixth Director of The Johns Hopkins University Applied Physics Laboratory on July 1, 1992. Dr. Smith was trained as a nuclear physicist and has spent his career at the Laboratory in a series of increasingly significant technical and managerial positions.

Dr. Smith came to the Laboratory in 1970, initially working in theoretical and experimental research relating to acoustic and non-acoustic detection of submerged submarines. He was elected to the Laboratory's Principal Professional Staff in 1978. In parallel with his important research achievements, he earned meaningful administrative responsibilities. In 1979, he became an Assistant Department Supervisor in the Submarine Technology Department, and in 1981 he assumed the additional role of Program Manager of the SSBN Security Technology Program. Three years later, he became Associate Department Supervisor. In 1988, he moved to the Director's Office, soon thereafter becoming Assistant Director for Research and Exploratory Development, then Assistant Director for Research and Programs, Associate Director in January of 1991, and Director in July of 1992.

Dr. Smith is active in numerous professional societies and has been involved as well in the Continuing Professional Programs of The Johns Hopkins University Whiting School of Engineering. He served as Chairman of the Joint DARPA/SSPO Task Force on Magnetohydrodynamic Phenomena in 1980-1981, he has served on the Editorial Board of the Johns Hopkins APL Technical Digest, and he was elected to the JHU/APL Advisory Board in 1982 and in 1988. In 1989 and 1990, Dr. Smith served on the Navy's Quo Vadis II Panel. In 1991, he served as Chairman of the Navy's High Gain Initiative Critical Experiment Review Panel.

Dr. Smith has a strong commitment to community service. Currently he chairs the United Way Community Partnership of Howard County, is Vice President of the Board of Directors of Leadership Howard County, and is a board member of The Columbia Foundation, United Way of Central Maryland, and the Maryland Science Center. In addition, he serves as the Chairman of the Steering Committee for Leadership U. a youth leadership program recently initiated by Leadership Howard County. In 1994, he chaired the Howard County United Way Campaign and in 1996, he served as Co-Chair of the High Technology Division of the annual campaign for United Way of Central Maryland.

Dr. Smith was born in South Dakota and grew up in Illinois. Following four years in the Navy, he attended the University of California at Davis, where he received his Bachelor's, Master's, and Doctorate degrees in physics. He was awarded membership in Phi Beta Kappa, Phi Kappa Phi, and Sigma Xi and is included in American Men and Women of Science:. Following his work at UC-Davis, he became a National Research Council Postdoctoral Research Associate at the Naval Research Laboratory. Shortly after coming to the Applied Physics Laboratory, Dr. Smith and his family moved to Howard County, Maryland. He and his wife, the former Claire Valine, have three grown children and eleven grandchildren.

April 1997

THE JOHNS HOPKINS UNIVERSITY
APPLIED PHYSICS LABORATORY
FUNDING (ALL SOURCES)
$ Million

	GFY			
	94	**95**	**96**	**97 (Est.)**
Total...	**467**	**394**	**372**	**370**
NAVY	**268**	**226**	**226**	**219**
NON NAVY	**199**	**168**	**146**	**151**
BMDO (Ballistic Missile Defense Organization)	**48**	**43**	**61**	***56***
DARPA (Defense Adv Research Projects Agency)	***5***	***5***	***14***	***16***
AIR FORCE	***23***	***9***	***7***	***7***
ARMY	***12***	***8***	***7***	***9***
OTHER DOD (OASD OSD, DMSO, DOD,...)	***7***	***5***	***5***	***4***
NASA (National Aeronautics & Space Adm)	***86***	***84***	***38***	***46***
OTHER NON DOD (DOT, DOE, NSF, TREAS, ...)	***18***	***14***	***14***	***13***

Weapon Effects From A

High-Altitude Nuclear Detonation

With Emphasis On

High-Altitude Electromagnetic Effects

Dr. George W. Ullrich

Deputy Director

Defense Special Weapons Agency

July 16, 1997

Testimony Before

Subcommittee on Research and Development

House National Security Committee

INTRODUCTION

Mr. Chairman, I am Dr. George Ullrich, the Deputy Director at the Defense Special Weapons Agency in the Department of Defense. I appreciate the opportunity to appear before you today to discuss this important issue.

It is interesting to note that exactly 52 years ago to the day, the world's first nuclear device was exploded at Trinity site, located on an isolated stretch of New Mexico desert in what is now the White Sands Missile Range. Among the team who witnessed that momentous event was Enrico Fermi, nobel laureate and perhaps the most brilliant of the Manhattan Project physicists. It was said that he was probably the last man of the twentieth century who actually knew all of the physics of his day. I mention it because it was Enrico Fermi who, prior to the Trinity Event, first predicted that nuclear explosions were capable of generating strong electromagnetic fields. Since then we have learned a great deal more about nuclear-induced electromagnetic phenomena and, in particular, about the phenomenon of high altitude Electro-Magnetic Pulse, commonly called "EMP."

The most common perception of a nuclear detonation is that represented by a mushroom cloud -- a burst at or near the surface of the earth. Such a burst results in a variety of weapons effects, most prominently blast and thermal, whose extent can reach up to several miles from ground zero, depending on yield. The only exception is radioactive fallout from a surface burst, which at low levels can traverse the entire globe. A high altitude burst, detonated at heights ranging from 50 to several hundreds of kilometers above the earth's surface, is also capable of generating a wide variety of effects and disturbed environments, the most far-reaching being EMP. Depending primarily on the burst height and to a lesser extent on yield, a high altitude burst can bathe a continental size region in EMP.

Such a detonation causes particular concern because of the sensitivity of modern electronics to strong electromagnetic fields. A knowledgeable adversary could attempt to exploit such a perceived weakness, thereby severely degrading the U.S. technological advantage, and he could do so in a way that would not likely provoke an immediate nuclear retaliation

A less well known effect of high altitude bursts, but also one with potentially devastating consequences, is the artificial "pumping" of the Van Allen belt with large numbers of electrons. The bomb-induced electrons will remain trapped in these belts for periods exceeding one year. All unhardened satellites traversing these belts in low earth orbit could demise in a matter of days to weeks following even one high altitude burst.

The United States' national military strategy is based, in significant part, on our technological advantages in such fields as electronics and computers. These are the enabling technologies for achieving Information Dominance, which contributed to our success in the Gulf War and will be vital on future battlefields. As outlined in A National Security Strategy for a New Century, The White House, May 1997, our national military strategy also emphasizes the importance of responding to asymmetries -- that is, unconventional approaches that avoid or undermine our strengths while exploiting our vulnerabilities. To quote from the report, "Because of our dominance in the conventional military arena, adversaries who challenge the United States are likely to do so using asymmetric means...such as WMD..." To preserve our technological advantage, DoD develops radiation hardened systems and tests them to assure survivability.

However, due to size and power reductions, modern electronics are inherently more vulnerable to some of the effects produced by a nuclear detonation. And each new generation, smaller and needing less power, exacerbates these vulnerabilities. Furthermore, as we make greater use of more affordable commercial parts and components, we potentially introduce new vulnerabilities into our military systems. Additionally, the military's increasing reliance on commercial space-based systems makes it more vulnerable to the nuclear weapon effects being discussed.

In my presentation today, I will provide a brief overview of the effects produced by nuclear weapons, to include lessons learned during both the United States' and Soviet Union's atmospheric nuclear test programs. Particular emphasis will be given to the most significant effects in a scenario in which an adversary uses one or a few nuclear weapons detonated at a high altitude. I will discuss what we have learned about providing affordable protection. Finally, I will mention what we do to simulate these threat level environments and how we perform testing to validate EMP hardness.

I should also note that the programs I will discuss are components within a broader set of Defense Department activities directed at sustainment of critical DoD nuclear mission competencies. These activities are described in detail in -- the May 1997 report by the Secretary of Defense on <u>Nuclear Weapon Systems Sustainment Programs</u> previously delivered to the Committee.

HIGH ALTITUDE NUCLEAR DETONATION EFFECTS

Based on over a half-century of research, we have developed an understanding of the effects produced by nuclear weapons. Since the end of the Cold War, we have added to our knowledge based on an analysis of information made available from the Soviet Union's nuclear test programs.

To understand the military consequences that can result from the high altitude detonation of even a single nuclear weapon, I will address:

- High Altitude EMP (or HEMP)
- System Generated EMP (SGEMP) and
- other Radiation Effects.

In keeping with your request, Mr. Chairman, I will direct most of my remarks to the topic of high altitude EMP.

High Altitude EMP

A nuclear weapon detonated at high altitude releases some of its energy in the form of gamma rays. These gamma rays collide with air molecules and produce what are called Compton electrons. The Compton electrons, in turn, interact with the earth's magnetic field, producing an intense electromagnetic pulse that propagates downward to the earth's surface. The initial gamma rays and resultant EMP move with the speed of light. The effects encompass an area along the line of sight from the detonation to the earth's horizon. Any system within view of the detonation will experience some level of EMP. For example, if a high-yield weapon were to be detonated 400 kilometers (250 miles) above the United States, nearly the entire contiguous 48 states would be within the line-of-sight. The frequency range of the pulse is enormously wide -- from below one hertz to one gigahertz. Peak electric fields can reach tens of thousands of volts per meter. All types of modern electronics are potentially at risk, from Boston to Los Angeles; from Chicago to New Orleans.

One of our earliest experiences with HEMP dates back to the resumption of atmospheric nuclear testing in 1962 following a three year testing moratorium.

Starfish Prime, a 1.4 megaton device, was detonated at an altitude of 400 kilometers over Johnston Island. Failures of electronic systems resulted in Hawaii, 1,300 kilometers away from the detonation. Street lights and fuzes failed on Oahu and telephone service was disrupted on the island of Kauai. Subsequent tests with lower yield devices produced electronic upsets on an instrumentation aircraft that was approximately 300 kilometers away from the detonations.

Soviet scientists had similar experiences during their atmospheric test program. In one test, all protective devices in overhead communications lines were damaged at distances out to 500 kilometers; the same event saw a 1,000 kilometer segment of power line shut down by these effects. Failures in transmission lines, breakdowns of power supplies, and communications outages were wide-spread.

System Generated EMP

When gamma and x-rays from a high altitude detonation encounter a satellite in space they excite and release electrons as they penetrate the interior of the system. This phenomena is referred to as system generated electromagnetic pulse (SGEMP) because the accelerated electrons create electromagnetic transients. Systems must be configured with special cables, aperture protection, grounding, and insulating materials in order to survive these transients.

SGEMP impacts space system electronics in three ways. First, x-rays arriving at the spacecraft skin cause an accumulation of electrons there. The electron charge, which is not uniformly distributed on the skin, causes current to flow on the outside of the system. These currents can penetrate into the interior through various apertures, as well as into and through the solar cell power transmission system. Secondly, x-rays can also penetrate the skin to produce electrons on the interior walls of the various compartments. The resulting interior electron currents generate cavity electromagnetic fields that induce voltages on the associated electronics which produce spurious currents that can cause upset or burnout of these systems. Finally, x-rays can produce electrons that find their way directly into signal and power cables to cause extraneous cable currents. These currents are also propagated through the satellite wiring harness.

Other Radiation Effects

A high-altitude detonation presents a double radiation threat to space based assets. Systems not protected by the Earth's shadow are exposed to the direct weapon outputs (gamma rays, x-rays, neutrons) and can be upset or damaged immediately if their range from the weapon is such that the radiation environments exceed electronic device tolerance levels. The second threat comes from the weapon-produced electrons that enhance the earth's natural Van Allen radiation belts. Satellites that repeatedly transit these enhanced radiation belts in their orbits will eventually exceed their total radiation dose tolerance and will degrade, then fail.

Weapon debris carries a significant percentage of the energy of the detonation and this radioactive material releases enormous numbers of high energy electrons through beta decay. This phenomena creates an artificial "trapped electron" radiation belt. The size and intensity of the belt is highly dependent on the yield, altitude, and latitude of the detonation. The energies of the weapon-induced trapped electrons are significantly higher than those of the natural environment. For example, a 50 kiloton (KT) weapon detonated at a 120 km altitude (75 miles) can produce electron densities several orders of magnitude higher than the natural electron environment in low earth orbit. These elevated electron densities can last for months to years and significantly increase the total ionizing dose accumulated by space assets that transit these belts. This increase in total dose accumulation can dramatically shorten the lifetime of satellite systems. Projected lifetimes of up to ten years can be reduced to a mere two months after such an event.

AFFORDABLE HEMP PROTECTION

We understand how to provide effective protection against EMP effects. The basic approach is to provide a shield that prevents damaging electrical pulses from entering a system. This requires protection at all electrical and mechanical penetrations. EMP hardening protocols have been published in standard handbooks and computer programs have been developed to facilitate system hardness designs.

EMP protection is also affordable. If accomplished during the design phase, the cost of EMP protection is a small fraction, one-to-five-percent, of overall system development costs. Done after the fact, when the unprotected system has already been fielded, it can be significantly more expensive.

To contribute to cost savings, we have an effort underway to develop integrated hardening methodologies that provide protection against multiple hazards. Our initial work focuses on integrated protection against the effects of both high altitude EMP and high powered microwaves produced with non-nuclear sources.

SIMULATION TO VALIDATE HEMP PROTECTION

We acquired much of our understanding of high altitude EMP effects and the protection needed from the development and effective use of nuclear weapon effects simulators. DoD currently operates a suite of simulators that provides the needed capabilities for large area, threat-level field illumination, direct current injection techniques, and low-level, continuous wave (CW) illumination to evaluate shield integrity and energy coupling efficiencies. These simulators are used in combination to validate a system's overall EMP protection.

STATE OF UNDERSTANDING

High Altitude EMP, System Generated EMP, and Radiation Effects are genuine, widespread hazards produced by even one nuclear weapon. We know how to protect against these EMP and radiation threats. Such protection is affordable, if provided for at an early stage in system design and development. For a tactical system, the cost can be as little as 1% of the total development investment; for strategic systems, a target of 5% is reasonable. Retrofitting protection after a system has been deployed can be considerably more expensive.

The pace of new developments in the fields of electronics and computers can be daunting. There is a new generation of microelectronics technology every eighteen months. Some of these new technologies are inherently more susceptible to nuclear threats. DoD has recognized and responded to these and other challenges. As outlined in the Secretary's May 1997 report on Nuclear Weapon Systems Sustainment Programs, additional funds have been programmed to ensure that core DoD requirements for advanced radiation hardened microelectronics technology are met. More recently, a Radiation Hard Oversight Council was established to ensure these efforts have appropriate visibility and oversight.

EMP does not distinguish between military and civilian systems. Unhardended systems, such as commercial power grids, telecommunications networks, and computing systems, remain vulnerable to widespread outages and upsets due to HEMP. While DoD hardens assets it deems vital, no comparable civil program exists. Thus, the detonation of one or a few high-altitude nuclear weapons could result in devastating problems for the entire U.S. commercial infrastructure. Some detailed network analyses of critical civil systems would be useful to better understand the magnitude of the problem and define possible solution paths.

CONCLUSION

Mr. Chairman, this completes my statement. I would be pleased to respond to the committee's questions.

STATEMENT OF ROBERT T. MARSH

CHAIRMAN

PRESIDENT'S COMMISSION ON CRITICAL

INFRASTRUCTURE PROTECTION

BEFORE THE SUBCOMMITTEE ON MILITARY

RESEARCH AND DEVELOPMENT

OF THE

COMMITTEE ON NATIONAL SECURITY

HOUSE OF REPRESENTATIVES

JULY 16, 1997

Good morning, Mr. Chairman and members of the Subcommittee. My name is Robert T. Marsh and I am Chairman of the President's Commission on Critical Infrastructure Protection (the Commission). I thank you for the opportunity to present my views about the potential of electromagnetic pulse (EMP) effects as a threat to our critical infrastructures.

My perspective arises from my service on the Commission, established by Executive Order 13010 on July 15, 1996. This is a joint government and industry commission charged with assessing threats to our critical infrastructures and their vulnerabilities. The President identified eight infrastructures as our national life support system. They are telecommunications, electric power systems, oil and gas transportation and storage, banking and finance, transportation, water supply systems, emergency services (such as medical, police, fire and rescue), and continuity of government services. The first line of the Executive Order says it all: "Certain national infrastructures are so vital that their incapacity or destruction would have a debilitating impact on the defense or economic security of the United States."

Our mission is to:

- assess vulnerabilities and threats to the critical infrastructures
- identify relevant legal and policy issues, and assess how they should be addressed
- recommend to the President a national policy and implementation strategy for protecting critical infrastructures
- and propose any necessary statutory or regulatory changes.

There are two categories of threats to our infrastructures: physical threats to tangible property and threats of electronic, radio-frequency or computer-based attacks on the information or communication components of critical infrastructures. The Commission's report to the President, scheduled for mid-October, 1997, will propose a national policy and implementation strategy to protect these infrastructures from both types of threats and assure their continued operation.

The EMP effects of nuclear weapons were thoroughly studied and well understood during the Cold War. At great cost, we hardened our critical command and control systems against such effects. We built extensive special test facilities and tested our C^3 systems to assure their continued operation under attack.

Obviously, the nuclear threat from hostile nations cannot be dismissed today, but we consider it a remote possibility. Likewise, we consider a terrorist acquiring a nuclear weapon and positioning it at the high altitude necessary for generation of an EMP burst that would debilitate our infrastructures to be a very remote possibility. Consequently, we are not considering any special measures to counter such a threat, though a high altitude EMP attack could devastate the telecommunications, power, and computational infrastructures.

We have also looked at localized radio frequency (RF) threat innovations and discussed them with RF weapons experts. It is theoretically possible to develop such weapons, but to my knowledge the practicality has not been demonstrated. Even if perfected, RF weapons would be targeted at local installations with limited, local effect. Nevertheless, progress in this area should be continually and actively assessed. If developments mature as some predict, we will have to address defensive measures. There is much promise in this technology, but today, I do not see any evidence that suggests capabilities seriously threatening our critical infrastructures.

The present likelihood of a terrorist obtaining a nuclear weapon is uncertain. But even if it happened, generating the high altitude explosion required to produce a devastating EMP attack would be extremely challenging. There are many easier, less costly ways to use nuclear weapons than delivery to a high altitude. Such an event is so unlikely and difficult to achieve that it does not warrant serious discussion at this time. The

Administration's policy is to prevent proliferation and unauthorized access.

In conclusion, Mr. Chairman, I believe the threat of a major debilitating EMP attack generated by a nuclear weapon is remote at this time. This is also true of the more localized effects of RF weapons, although this area needs to be kept under surveillance and may warrant the development of countermeasures in the future.

This completes my statement, Mr. Chairman. I will be pleased to answer any questions you or Subcommittee members may have.

Statement of

The Acting Deputy Under Secretary of Defense for Space

Mr. Gil I. Klinger

Before

the

Subcommittee on Military Research and Development

of the

House Committee on National Security

16 July 1997

Not for Publication

Until Released by the

House Committee on National Security

Introduction

Mr. Chairman and members of the Committee, it is an honor for me to appear before this Committee to address the potential threat to our national security satellites posed by electromagnetic pulse (EMP). Before I address my remarks to the topic of EMP, I would like to give you a brief background on the Deputy Under Secretary of Defense (Space) (DUSD(Space)) and our responsibilities. This will assist in explaining the scope of our activities.

DUSD(Space) is the Secretary's principal staff assistant for space matters. We are developing the plans and architectures which will effectively lead us through the next decade. In this capacity we develop, coordinate and oversee the implementation of DoD policies and guidance for military and intelligence space activities; oversee the implementation of DoD space systems related strategy and plans; and, oversee DoD space mission and system architectures, acquisition and technology programs. We are key in analyzing and implementing the National Security Section of the National Space Policy and other Presidential space policies. We successfully lead the Overarching Integrated Product Team (OIPT) for the Space-Based Infra-Red System keeping this major acquisition program healthy and on track. We are currently leading OIPTs on the Evolved Expendable Launch Vehicle (EELV) and the

National Polar Orbiting Environmental Satellite System (NPOESS). In accomplishing these tasks we coordinate and monitor the implementation of policies established by other DoD offices and agencies. For example, electromagnetic pulse survivability policy and standards are established by the Office of the Assistant to the Secretary of Defense (Nuclear, Chemical, Biological Defense Programs). These policies and standards are included in space systems design and acquisition appropriate to the validated threat to the system.

Hardening guidance is established by national and defense policy and the Joint Staff. Space systems are hardened consistent with this guidance. We recognize the growth in the use of space systems as articulated in Joint Vision 2010. In light of this potential massive use of space capabilities, both national security and commercial systems will require a review and study of critical infrastructure components. DUSD(Space) is heading this DoD-wide study.

In the past, National Security Space Systems have been hardened to varying degrees. This variation continues today. Hardening all systems to the maximum level incurs heavy cost penalties. We are studying a range of protection measures which will enhance the entire satellite constellation. This review will look at these protection measures with cost as an independent variable. The Joint Space Management Board (JSMB) received and approved a comprehensive space protection study conducted by the DoD Space Architect and the National Reconnaissance Office. Implementation of the recommendations contained in this study of current and projected issues is being planned.

Electromagnetic Pulse Effects on Space Systems

As Dr. Ullrich of the Defense Special Weapons Agency has explained, the most common usage of the term electromagnetic pulse (EMP) refers to the complex radio frequency waveform generated by a nuclear device being detonated in the atmosphere (also known as High Altitude EMP, HEMP). This "transmission" can contain large amounts of energy that can be "received" by a military or civilian system in unexpected ways that damage electrical components. The process is somewhat similar to the static one picks up on an AM car radio during a thunderstorm due to lightning. The amount of HEMP energy that "couples" to the system depends on many factors, specific system design and distance from the source being among the most important. The car radio analogy is still appropriate here with an FM radio being less disturbed and very distant lightning producing no static. Spacecraft are far from a HEMP event and are not damaged. The effects of

HEMP on satellite ground stations can be more disruptive, and depend on a number of factors (e.g., altitude and distance of burst or design of ground station). Military communications facilities such as the National Military Command Center have HEMP hardened facilities.

Another "form" of electromagnetic pulse previously discussed is system generated EMP or SGEMP. This effect happens only in a vacuum and when X-rays produced by a nuclear device strike a satellite and displace electrons throughout the spacecraft. The movement of these electrons produces large currents in the satellite's structure and electrical systems usually with detrimental effects. An analogy would be plugging a nine-volt transistor radio in a household electrical socket. The amount of SGEMP generated in a spacecraft is dependent upon the nuclear weapon's yield, X-ray spectrum, distance to the nuclear device and, most importantly, the design of the satellite. The design and testing of military space systems for vulnerability to SGEMP effects is the responsibility of the individual program offices. The adequacy of nuclear survivability features is an item that is considered by the Defense Acquisition Board. As part of our ongoing space protection efforts, we will be looking explicitly at the issue with cost as an independent variable.

Radiation Hardened Microelectronics

The "radiation effects," discussed by Dr. Ullrich, are the result of natural and man-made sources. Natural sources are cosmic rays, charged particles trapped in the Van Allen belts and solar flares. Engineers design commercial and civil satellite electronics to withstand this naturally occurring radiation. DoD satellite systems must be designed to survive much more severe radiation environments induced either by hostile actions (JCS Level 1 hardening requirements) or high ambient radiation orbits (Global Positioning System). The specially designed and manufactured electronic components are termed to be "radiation hardened" or simply "Rad Hard." The availability of Rad Hard electronic components is crucial to survivable DoD satellite systems and receives special attention from DUSD(Space).

In December 1996 an Integrated Product Team (IPT) chaired by the Assistant Deputy Under Secretary of Defense (Space Systems Acquisition) reported to the Under Secretary of Defense (Acquisition and Technology) on the industrial base that designs and manufactures Rad Hard components. The IPT found that:

- There is an insufficient commercial business base to support a Rad Hard industrial capability, the government, primarily DoD, is the principal customer in this "niche market,"
- Government investment in advanced radiation hardening technology is adequate but there is insufficient funding for transitioning these technologies to production
- Government and industry competency has decreased as the industry down sizes.

The IPT's Rad Hard strategy was to generate economies of scale by providing the "seed money" for Rad Hard parts development and production thereby assisting the manufacturers to remain in the business while they develop a non-DoD market. The IPT specifically recommended:

- Establishing a DoD-level group to oversee and coordinate DoD investment in Rad Hard electronics with a companion interagency coordination group
- Funding an annual Rad Hard investment program at $60 - 70 million
- Creating a graduate-level initiative to train more electrical engineers in radiation hardening technology

In May 1997 the USD(A&T) approved the IPT's recommendations and directed a DoD wide initiative implementing them. Detailed actions were assigned to DoD staff offices and to the Air Force. These tasks are currently being accomplished with the establishment of the Radiation Hardened Electronics Oversight Council being the highest priority. We will update the committee on the implementation of the DoD Rad Hard initiative and the progress in creating an interagency coordination mechanism.

These issues become more important as DoD transitions to dependence upon commercial satellite systems to provide cost effective ways of meeting mission requirements. We are studying this issue in conjunction with US Space Command and the NRO. The vulnerabilities of both National Security and commercial assets will have to be taken into account. To the extent we migrate critical space functions to commercial providers, we need to focus on protection and negotiate appropriate hardness levels.

Mr. Chairman, this completes my statement. I would be pleased to respond to the committee's questions.

45-989 98 - 4

EMBARGOED UNTIL RELEASED BY THE SUBCOMMITTEE

PREPARED STATEMENT

Supporting Invited Testimony To Be Presented By

Dr. Lowell Wood*

On

Electromagnetic Pulse (EMP)

Before The

Subcommittee on Research and Development

Committee on National Security
United States House of Representatives

10:00 AM, 16 July 1997
2118 Rayburn House Office Building

"If you would have peace, prepare for war."
– Benjamin Franklin

I am grateful for the Committee's kind invitation to offer testimony on electromagnetic pulse (EMP) and its implications for our Nation's military capabilities and for the continuation of American civilization.

BACKGROUND. I have been an interested observer of both American and foreign capabilities with respect to electromagnetic pulse (EMP) phenomena for three decades, and I have been involved with both offensive and defensive aspects of electromagnetic pulse weaponry for more than two decades. During the '70s, I served on the Defense Nuclear Agency's Scientific Advisory Group on Effects (SAGE), the DoD's senior technical review group concerned with electromagnetic pulse, as well as all other military nuclear issues having a technical character. In the late '70s and early '80s, I worked on "Third Generation" nuclear weaponry, a major component of which was nuclear explosive-driven generators of electromagnetic pulses of potentially greatly increased efficiency and military effectiveness; spinoffs involving non-nuclear means of generating potent EMP also engaged my attention. Later in the '80s and early '90s when strategic defense was emphasized, I worked on the development of defensive technologies of very high efficacy against nuclear EMP, with particular reference to military space systems. With the fall of the Soviet Union, my attention has turned to the implications of EMP in a nuclear-multipolar world, while remaining mindful of the EMP implications of the enduring Russian nuclear force structure.

*** Visiting Fellow, Hoover Institution on War, Revolution and Peace, Stanford University, Stanford CA 94305-6010, and Member, Director's Technical Staff, University of California Lawrence Livermore National Laboratory, Livermore, CA 94550. Opinions expressed herein are those of the author only. House Rule XI, Clause 2(g)-mandated information is appended.**

I have been privileged to appear on a number of occasions before this Subcommittee and its siblings during the past two decades, testifying on a variety of national security topics. I have also served the Committee in a technical consulting capacity, initially under Chairman Les Aspin nearly a decade ago, and more recently under the leadership of Chairmen Floyd Spence and Duncan Hunter.

SOME HISTORICAL PERSPECTIVE. More than a third-century ago, due both to commentary from our British allies and to some truly striking experimental results, military technologists in the United States became generally aware that *high-altitude* nuclear explosions often generated electromagnetic effects of completely unprecedented magnitudes, physical and temporal scales – and effects on both the physical environment and human handiwork. (It had been appreciated in a rather qualitative manner for some time previously by American workers that electromagnetic phenomena of singularly large magnitudes and quite exotic natures occurred in the immediate vicinity of nuclear fireballs created near the Earth's surface, but these effects were largely ignored against the background of the nuclear explosion-unique blast and heat effects.)

The first American high-altitude nuclear weaponry experiments after the Soviet breaking of the nuclear test moratorium of '58-'61 revealed a wealth of phenomenology of completely unprecedented – and largely completely unanticipated – character. Most fortunately, these tests took place over Johnston Island in the mid-Pacific rather than the Nevada Test Site, or "electromagnetic pulse" would still be indelibly imprinted in the minds of the citizenry of the western U.S., as well as in the history books. As it was, significant damage was done to both civilian and military electrical systems throughout the Hawaiian Islands, over 800 miles away from ground zero. The origin and nature of this damage was successfully obscured at the time – aided by its mysterious character and the essentially incredible truth.

Intensive effects commenced to understand what had happened – and what might happen in the event of hostilities involving high-altitude nuclear weaponry usage. These efforts were spurred by the knowledge that the Soviets had experimented extensively with high-altitude nuclear weaponry, including some uniquely high-yield explosions, during their '61-'62 test series, and presumably understood the implications of these at least as well as we did. American efforts were complicated by the cessation of high-altitude testing associated with implementation of the Atmospheric Test Ban Treaty in '63, so that access to experimental truth was greatly complicated and, in some crucial respects, entirely precluded.

At this point, the Soviet Union and the United States commenced to engage the nuclear EMP issue somewhat analogously to two men fencing with very sharp blades in utter darkness: both knew that the weaponry which they wielded was extremely potent, but neither knew the other's time-varying posture, let alone the precise location of either vulnerable spots or especially well-armored ones. This deadly duel continued for three decades, through the collapse of the Soviet Union. It continues today. It will continue into the foreseeable future. ***EMP crippling of the American military machine and of the modern American nation remain real prospects.***

THE NATURE OF EMP. Electromagnetic pulse is sufficiently alien to ordinary human experience that it seems to many to be either magic or illusory. Such entirely understandable human reactions have not facilitated the development or implementation of apt responses to the profound threats which it poses, either by political or military leaders. Thus, a few operationally-oriented fundamentals may be of use:

EMP is really severe static electricity, everywhere, all at once. Without needing to understand the undeniably esoteric means by which EMP arises in various military circumstances, it suffices to recall that it presents itself as something closely akin to static electricity, extremely intense but exceedingly brief, everyone within line-of-sight to a high-altitude nuclear explosion, "all at once." (This "static electricity" pulse is carried on radio-frequency electromagnetic waves of uniquely high intensity. The bomb's extraordinarily intense prompt radiations essentially transform the entire atmosphere underneath it into a gigantic radio transmitter-antenna – for a very brief interval.)

Extended metallic structures within line-of-sight of the explosion – telephone and electrical lines, radio and TV antennae, fence wires, etc. – then serve to gather up the broadcast energy of EMP and deliver it into whatever they connect to, often with locally ruinous results which are basically similar to those resulting from a lightning-strike. Since it travels at the speed-of-light, EMP arrives essentially instantaneously, from the general direction of the explosion.

EMP can blanket an entire U.S.-sized continent from a single source. EMP originates in the interaction of gamma-radiations from a nuclear explosion with the Earth's atmosphere at altitudes of a few dozen kilometers and propagates predominantly toward the Earth's surface. Thus, since you can readily see a bomb explosion a few hundred kilometers above the central U.S. from anywhere in the "lower 48", the EMP arising from that explosion's interactions with the Earth's atmosphere can also "see" you.

To be sure, at greater distances, the intensity of the pulse will be smaller, but not necessarily as indicated by the familiar inverse-square-of-the-distance law. Likewise, its severity is only roughly proportional to the yield, or total energy production, of the bomb. (The sharpness of the EMP actually depends rather sensitively on esoteric aspects of the bomb's design and operation. Low-yield specially-designed bombs may pose as significant EMP threats as do high-yield "ordinary" ones.)

EMP doesn't linger. Since it arises from a nuclear explosion's promptly-emitted gamma radiation interacting with the Earth's atmosphere, nuclear EMP goes away very quickly. It is a phenomenon of real interest only for time-scales of microseconds – millionths of a second. (Within this time-frame, however, it can be quite compelling.) It has none of the character of nuclear radioactivity or fallout.

EMP isn't sensed by people, and it doesn't damage the human body. The nervous system and associated sensory faculties of people are essentially totally insensitive to electromagnetic radiation of the frequency and the time-duration of EMP. We don't sense it in any way. Because it arises and then goes away so exceedingly quickly, electrical currents due to it do not really begin to flow in our bodies, and no physiological damage of any kind takes place. EMP really "speaks" only to metallic objects, and to things connected to them.

EMP is much more threatening to big electrical systems than to small ones. Because metallic objects of many different shapes can effectively gather up and then concentrate the energy associated with EMP, physically large systems comprised of metal – lines, cables, wire and dish antennae – often manifest exceptionally great vulnerability to EMP damage. Their spatially extended components "harvest" the broadcast EMP energy, which falls fairly uniformly over wide areas, and bring it to wherever the system's "barn" may be – the often-centrally located components of the extended system which may be quite sensitive to electrical overload. Physically small systems usually don't get EMP "illumination" so well-collected or -focused within themselves, and thus tend to be more durable to its effects.

EMP is much more threatening to modern electronics than to old-fashioned ones. Older electrical and electronic systems are generally built out of massive components, which are innately much more tolerant of the effects of EMP. Vacuum tubes, for example, are extremely EMP-rugged, while the ever-tinier transistors which have almost totally replaced them in the U.S. military machine – as well as in U.S. civilian electrical/electronic systems of all types – are ever-more-vulnerable to EMP destruction. The Soviet technological lag behind the Americans has been a substantial – and vigorously exploited – advantage in this somewhat perverse respect.

EMP in space is different from EMP near the ground, and is typically nastier. EMP arising in spacecraft due to exposure to prompt nuclear hard-x-ray and gamma radiations – even from very great distances – is often tedious in the extreme to eliminate effectively and with adequately great assurance. (It is assuredly possible to accomplish, however, even against the most severe threats.)

Nuclear EMP correspondingly poses an extremely serious threat to the assured functional survivability of space assets, both military and civilian, the more so as the essential system-level testing always was quite expensive and currently is effectively impossible.

EMP defenses are simple, and traditionally have added ~10% to military system costs. For typical military systems which do not operate in space, the rule-of-thumb has been that robust hardening against EMP effects adds roughly 10% to the total system life-cycle cost, if such hardening is engineered-in from the outset.

Such hardening consists primarily of high-integrity albeit thin (e.g., tinfoil-like) metallic shielding to keep the EMP radiation out of enclosed volumes containing vulnerable systems components <u>and</u> of special electrical devices – e.g., high-tech lightning arresters – for protecting absolutely essential penetrations of such metallic enclosures from inadvertently admitting significant amounts of EMP energy into the interior "sanctuary." The major fractions of this ~10% added-cost have traditionally been spent in performance-testing and *life-cycle maintenance* of EMP-robustness, not in creating the "sanctuary" itself. Significant economies might be realized in these cost-dominating areas in future efforts by intelligent use of more modern technologies. If, on the other hand, EMP hardening is back-fitted to an existing military system, costs have generally been in the range of 20% of total system cost.

SOME FUNDAMENTAL TECHNICO-MILITARY DIFFERENCES. There were several fundamental differences in the technical and military approaches which the Soviet Union and the United States took toward EMP. These differences are reflected in the postures of the two nations' military machines today, and likely will be enduring ones.

The Soviets basically decided that EMP represented not only an exceptionally severe threat to the integrity of their military apparatus and their civilian infrastructure, but also offered extraordinary opportunities to their strategic offensive forces. Relatively deficient in computational modeling tools with which to understand the quantitative details of EMP generation and interaction with a wide variety of particular structures and systems, they took a generic, highly empirical "belt and suspenders" approach to protection of both military and civilian systems against EMP, deploying protective hardware quite lavishly (as compared to the U.S.) and providing extensive counter-EMP training to both civilian and military personnel involved in the operation and maintenance of these systems.

Soviet strategic strike forces characteristically featured weaponry very well-suited to efficient EMP generation over exceptionally wide areas. ***That EMP strike component exists today in the Russian strategic order-of-battle, moreover at its maximum Cold War strength.*** *I very confidently predict that it will be one of the last features of Soviet strategic nuclear weaponry to be retired from the Russian force structure.* It has long be considered highly likely by U.S. strategic war planners that a Soviet first-strike would commence with a multi-explosion "laydown" of high-intensity EMP all over the continental U.S., significantly before any target on the ground is brought under attack, simply because the cost of such an attack-commencement is low and the benefits gained are great. Indeed, memories of strategic war games long past have as a major common feature the beginning of the game with a massive Soviet EMP laydown all over the U.S., followed immediately by an extended time-out while the game's referees rip up huge handfuls of U.S. military capability of all types and throw it away as EMP-ruined.

We Americans, in contrast, collectively saw EMP as a major nuisance which could be rather precisely understood, defended against "good enough" – and thereafter largely ignored. The Defense Atomic Support Agency (DASA), succeeded by the Defense Nuclear Agency (DNA) and now by the Defense Special Weapons Agency (DSWA), in exceptionally fruitful long-term collaboration with dedicated components of American industry (of which the RAND Corporation Physics Department, later re-organized as R&D Associates, and the Mission Research Corporation were particularly distinguished leaders), developed a really outstanding technical appreciation of EMP, how to model and simulate it with high fidelity, and how to effectively defend major military systems against it.

Indeed, I estimate that half of DASA/DNA/DSWA's third-billion dollar ('96 $) time-averaged annual budget was expended for such purposes for three decades.

Programs then came into existence to express and embed this evolving understanding – excellent albeit imperfect – of EMP in major American strategic warfare systems, primarily the offensive ones but also the defensive components. However, because neither supercrats nor senior commanders really understood – or, in some cases, believed in the existence of – EMP and its effects, these EMP hardening programs generally followed uncertain trumpets, and their average effectiveness was not exceedingly high. (At that, U.S. strategic military systems were vastly better EMP-protected, on the average, than were our tactical ones.)

Some CINCs stand out in my memory as exceptionally diligent in their efforts, the results of which were especially praiseworthy. (A few senior Navy admirals, enjoying unusually great tenure and discretion over the resources of their large commands, did very well by the enduring National interest in these respects.) Too often, though, protecting against a poorly-understood, deemed-unlikely threat of a semi-magical character lost out in the unceasing battles-for-resources, and was deferred, largely or completely, to "next year" – a well-known interval of time which never quite arrives in DoD-Land. In some notable EMP-hardening programs, sustained and strenuous efforts were made without securing desired results, outcomes which were sometimes obscured to the present day by lack-of-candor leveraged with high security classifications. Case histories abound, but are not appropriate for open discussion.

As a result, the present-day U.S. strategic force structure is a veritable patchwork quilt with respect to its EMP durability. The bottom line is that, in really bad weather, this quilt won't keep at all warm the fundamental National interest. *This situation is undoubtedly well-known, even in many of its details, to our potential near-peer adversaries – and it presumably incentivizes exploitation-directed efforts on their part.* ***At that, America's strategic forces are much better-postured against EMP attack than are our day-to-day, tactical military forces.***

The EMP robustness of the civilian infrastructure of the United States can be summarized far less equivocally: it is entirely non-existent. *Our civilian telephony, electricity, communications and electronics plants are all naked to our nuclear-armed enemies.* ***Even a modest, single-explosion EMP attack on the U.S. would likely devastate us as a modern, post-industrial nation.***

PECULIAR ASPECTS OF EMP ATTACKS. Indeed, a nuclear EMP attack on a nation is, in the large, the *obverse* of what the neutron bomb was asserted (utterly falsely by anti-deployment-directed Communist propaganda, but nonetheless with great political effect) to be in the small: an arch-capitalist weapon which killed people but didn't destroy the capital plant in which the people were located. *EMP weaponry (potentially even in single copy), in acute contrast to this now-ancient canard, potentially destroys in a highly effective manner the high technology electrical/electronic plant of any advanced nation – the heartland of modern civilization – while not directly harming people at all.*

It is profoundly unsettling that the electrical/electronic infrastructure of a large modern nation – which may be valued at more than ten thousand dollars per capita, or a few *trillions* of dollars for a nation such as the U.S. – can be so threatened from afar by a single nuclear explosion. That this can be done without harming people – potentially even invisibly, if done in broad daylight – gives real pause for thought, in a still-troubled, nuclear-multipolar world.

ASPECTS OF THE INTERNATIONAL STATE-OF-PLAY. Several aspects of the current and likely-future geopolitical state-of-play seem impacted by such considerations.

Through the end of the Cold War, we Americans could "attribute" any EMP attack on us with exceedingly high confidence to precisely one source: the Soviet Union. Moreover, we usually anticipated that such an attack would merely comprise the precursor of an all-out attack and, as such, "will be met with a full retaliatory response." Toward the end of the Cold War, American strategic war planners worried about more nuanced Soviet attacks, possibly EMP-intensive ones involving quite limited damage-on-the-ground, and how to most appropriately respond to such damage-intensive but "casualty-poor" attacks. Such perplexities seemed largely obviated by the end of the Cold War and the cessation of such "virtual hostilities" with the Soviets.

But were they really? It is widely-known that we Americans contemplated, briefly and in a non-pervasive fashion, a nuclear EMP laydown on Iraq (a NPT signatory legally entitled to immunity from all nuclear attacks) as an exceptionally high-effectiveness commencement to Operation *Desert Storm* – and that two-thirds of the American people polled on the subject in that season supported the taking of such steps to protect the lives of American troops. It certainly should not be surprising if other nuclear-capable nations were thereby stimulated – if indeed any such external stimulus was needed – to contemplate employment of a similar tactic against their various adversaries, of which the U.S. may well be one.

What would the U.S. response be to a nuclear EMP "bolt from the blue" – or even one from a geopolitically overcast sky? What if such an attack, e.g., executed with a single rather modest Earth-orbiting bomb, arguably could have been mounted not only by Russia, but also by China or India or Iran – or North Korea? Particularly if none of our fellow citizens died as a direct-and-immediate result of such an attack, what degree of certitude of attack attribution would we require of ourselves before an American President would order a retaliatory strike imposing condign punishment on the suspect nation? Paralyzed as a modern nation, thrown back decades in time in industrial capabilities but still retaining a reasonably full set of nuclear teeth in our national mouth, how would we Americans then choose whom to bite – if anyone?

That scenarios of this general flavor are currently considered "within the pale" is illustrated by the "Army 2020" war game conducted at Carlisle Barracks earlier this year. Especially notable for its openness, it postulated a U.S. expeditionary force in the Ukraine clashing with an invading Russian force, two decades hence. When the Russian force fared poorly in ground combat operations, the Russian

General Staff used a set of nuclear explosions in space to effectively destroy the 'high eyes and ears" of the U.S. military – and all the civilian comsats and all the Russian space systems, as well – in order to express "national resolve." In addition to the far-distant Russian nuclear explosions giving American decision-makers real pause for thought, the entirely unexpected, abrupt and total loss of the "high ground" conferred by U.S. space assets nearly cost the American force its collective skin. Just as this game was ending in Russian-American armistice, the Chinese, noting America's unprecedented military incapacity, commenced to make their long-expected moves in the Far East. . . .

At that, wafting out of this unusually thought-provoking exercise was a faint aroma of "Blue-preferred Red responses," a well-known key ingredient of politicomilitary folly. The Army's game-designers were willing to postulate nuclear explosions in space of a flavor which acted over time-scales of hours to days to dramatically "burn down" American space assets largely owned-or-operated by the Air Force. However, they didn't care to consider an arguably equally plausible Russian nuclear EMP laydown over Ukrainian territory within which the American expeditionary force was operating – which, without inflicting casualties, likely would have devastated the electrical/electronic sinews of American tactical assets (ones incidentally almost entirely owned-and-operated by the Army).

Indeed, EMP laydowns constitute a generically attractive response on the part of any major regional power – not just Russia – to virtually any American power-projection attempt. They exemplify what is termed a "technologically asymmetric response" to the impending *Revolution in Military Affairs*, one in which our adversary acts purposefully to leverage his set-of-strengths and exploit our set-of-weaknesses. (Saddam Hussein fought us entirely on our terms; we must assume that we will not be gifted with a similarly inept adversary for some long time.) *Because a very small number – potentially just one – nuclear weapon exploded at high altitude over an American expeditionary force attempting forced entry against a major regional power could potentially tip the balance against our efforts, all such powers who contemplate someday possibly confronting us will be incentivized to develop, acquire or retain nuclear weaponry – quite contrary to the goals of ongoing nuclear nonproliferation efforts* _and_ *to the objectives of the Revolution in Military Affairs.* (It should be noted in this context that there are over 10,000 ballistic missiles presently owned by over 30 countries which are potentially capable of lofting a nuclear weapon to high altitudes over proximate U.S. forces – and that _none_ of the ballistic missile defense programs of the current Administration aim at military "products" which could defend against such "pre-apogee" attacks.)

Both as a demonstration-of-military capability and a show-of-national resolve, exploding a nuclear weapon continues to have no peer. (The South African example naturally comes to mind in the current context, both with respect to its motivations and its successful covertness.) If exploded so as to also cripple opposing military forces without also inflicting mass casualties, the potential attractiveness of such weaponry likely becomes compelling. A few nuclear weapons and unstoppable delivery systems (e.g., ballistic missiles facing only Clintonesque missile defenses) which can throw them into space, one at a time, over an invader's forces thus naturally rise to the top of the "wish list" of many types of national leader.

Thus, for several reasons, each one good-and-sufficient, the U.S. would be well-advised to manifest far more effective concern than prevails at present regarding EMP attacks against its national territory and against its forces abroad. Conventional approaches to threat assessment – i.e., those which attempt to conservatively assess both capability and intent – may lead to Pearl Harbor-class catastrophes in the context of EMP attacks.

EMP ISSUES FACING THE DEFENSE DEPARTMENT. Against such current and anticipated-future geopolitical backgrounds, then, what are the major EMP-related issue-sets facing the DoD?

First, having incompletely triumphed in its EMP hardening efforts during the Cold War – when the threat was clear-and-present and the resources were relatively plentiful – how can credible DoD responses to present and emerging EMP challenges be rationally anticipated now? Specifically, how would a reasonable skeptic be persuaded of the seriousness and effectiveness of any new-found DoD sense-of-purpose with respect to EMP defenses?

Second, how can DoD reliably eliminate the prospect of a single moderate-to-large yield nuclear explosion occurring at high altitude over the U.S.? (Indeed, how can it detect that such an EMP attack is underway, or that it is likely?) How can it robustly attribute the origin of such an attack (noting that at least Russia, China and India manifestly have the capability to execute such an attack today)? *If reliable defense is not feasible and robust attribution-for-deterrence is not possible, how is eventual attack to be rationally judged to be at all unlikely?*

Third, how much is DoD willing to carve out of its present-day, shrinking-in-real-terms budget to defend itself and (a clearly independent issue) to defend its Nation-sponsor from EMP attacks? Where, in particular, is how much money going to come from? How is this amount to be seen as credible, relative to what level of EMP defensive/hardening requirements?

Fundamentally, DoD must decide that it is significantly more important to engage EMP defenses now that it was a decade or two ago, or precisely no intra-Departmental motion is possible in the current budgetary environment. (The emphasis given EMP issues in the recent QDR seems diagnostic in this respect.)

It must then learn the lessons of how-and-why it succeeded in some EMP defensive programs and more-or-less failed in many others, over the past three decades. (The associated large-scale scrapping of pleasant fairy tales and reversion to full candor necessary for such self-education in this area may be possible only because so much time has elapsed since most such programs culminated.)

Then resources adequate to the EMP defensive tasks must be identified and robustly fenced, and a single, highly-capable, long-term-accountable, senior-level-reporting individual given responsibility for all aspects of program execution, from start to finish. *Only* under such circumstances is it realistic to rationally expect real progress.

EMP ISSUES FACING THE CONGRESS. With all due respect, the only *fundamental* issue facing the Congress is determining the degree of its own concern regarding the EMP threat to National military capabilities and to the at-risk portions of the Nation's infrastructure.

Once the degree of this concern is determined, the Congress may then ascertain quite readily whether or not it is sufficient to elicit a voluntary, "If you want it done, we're willing to do it" response from the DoD. If it isn't, then the only remaining question is whether the Congress is minded to mandate in statute the desired response from an unwilling DoD – with all the well-known risk, cost and oversight hassles thereto pertaining.

RECOMMENDATIONS FOR CONGRESSIONAL CONSIDERATION. If the Congress chooses to initiate an EMP defensive program, I respectfully recommend that any such initiative include the following features:

Mandated Organization-For-Success. A brand-new, single-purpose organization is a political luxury in DoD, but it is a time-proven "high road" to programmatic success. Such an organization must be run by a highly competent, surpassingly dedicated "benevolent dictator" whose enjoys unquestioned tenure and direct reporting to top-level DoD officials. (E.g., the House's FY'98 Defense Authorization Bill mandates such a reporting line for the Director of BMDO).

The government staff of this organization must be exceedingly few in number, exceptional in professional preparation and highly empowered. Industrial collaborators must be carefully selected for across-the-board competence, trusted and empowered thereafter and without exception – and likely compensated on a CPFF basis. External meddling in organizational business must be sharply minimized, and programmatic turbulence of all types – particularly with respect to budget and "mission creep" – rigorously suppressed.

DoD doesn't create many of these operations, for reasons both regrettable and obvious; if the Congress really wants EMP defensive programmatic success, it'll mandate such an operation into existence. (All of these are features identified in a recent RAND study of the F-117 Stealth fighter program to be common to those DoD acquisition programs that perform in peacetime like the really crucial ones usually do in wartime.)

Mandated *Across-The-Board* Competition. All routes to programmatic success must be fairly and objectively evaluated, and defensive hardware from all vendors evaluated objectively on a common basis. Insisting on "picking winners and losers" is a regrettably common way for ego-blinded DoD program managers to fail. *Mandating* such pervasive competitive arrangements is the only reliable way to gain them.

Assured Managerial Accountability and Stability. Most DoD acquisition programs perform as abysmally as they do, relative to the closely comparable people-sets working in American industry, primarily because managerial stability is distinguished by its absence and managerial accountability is correspondingly non-existent. "State property is nobody's property" as the old Soviet saying went, and the Government's interest in programmatic success of its Defense acquisition programs is almost invariably "co-owned" by precisely no one, civilian or military. Even a superficial comparison of Soviet and American experience over the past few decades indicates clearly that, without some type of proprietorship, no "property" will be decently looked after, and the long-term consequences likely will be telling ones. The Congress would be well-advised to act accordingly with respect to creation of EMP defenses: the program's senior managers should be "lashed to the mast" until the programmatic ship weathers the inevitable storms.

Design-To-Cost Focus. One pervasive problem encountered in the DASA/DNA/DSWA EMP hardening program was a single JCS specification for EMP hardness which had to be met, no matter what the attendant cost or difficulty might be for particular military systems. While this hardness level had a fundamental, rational basis, it might be more appropriate under prevailing circumstances to have two or more hardness levels, and to design hardening to a specified, not-to-exceed total cost-per-system hardened. Some significant degree of hardening is always better than no hardening at all. Congressional cost caps of undoubted robustness wonderfully concentrate both the bureaucratic and the defense-industrial minds – particularly if profits-and-fees are specified-in-advance to be paid out the wrap-up portion of the program's capped total budget.

"Free Fixes" Arrangements With DSWA Customers. One of the standard ways-to-crash in the DASA/DNA hardening program arose from the fact that the agency's folks were chartered to provide all manner of free assessment, advice and technical consulting to the military customer, but the customer had to pay all of the actual costs of hardening his own systems. As a consequence, only the unusually rich or the exceptionally diligent customer ever managed to pay for more than a small fraction of the hardening work to be done. An obvious fix for this fatal hesitation is to subsidize – perhaps even completely – the hardening work, and maybe even offer incentives to sign up for hardening. Again, cost caps on fixes – perhaps supplemented by fee incentives for on-spec/under-budget performance – may serve usefully to keep overall program budgets on-track.

Independent Periodic Assessments. For nearly three thousand years, the applicable maxim of Western jurisprudence has been "No man is an apt judge of his own cause." Particularly in its oversight of <u>any</u> Congressional initiative – one in which DoD co-ownership may be somewhat lacking – the Congress would be well-advised to commission independent reviews and assessments of programmatic progress made and problems encountered. No honest program will object to a single swiftly-executed annual review by competent-and-objective folks who can be "brought up to speed" without undue effort or delay.

Frequent Full-Up Trials Of Prototypes. Those DASA/DNA EMP hardening programs which were more likely to be successful had frequent testing of hardened sub-systems and systems as a major feature. Such exercises build confidence that progress is being made while also uncovering problems when they're small and relatively quick-and-cheap to fix. They're not diversionary, and should be mandated if simple encouragement to do so is insufficient.

It is notable in this context that the system-level EMP hardness of most all American space assets currently is entirely conjectural, due to a pervasive lack of realistic full-up testing in recent years; any assertions to the contrary can be most charitably characterized as white lies. That a National capability to conduct realistic full-up system-level testing of critical space assets was developed and exercised so *tellingly* – and then abandoned – is a major "blot on the copybook" of several OSD crews with respect to safeguarding the National interest. A clear and firm Congressional mandate is the only "sure cure" for this problem.

Performance Certifications. Several large-and-venerable investment houses have sustained titanic losses in the past few years by allowing their high-rolling traders to run their own back-offices, with the result that losses were effectively concealed until they ran into ten figures. Government programs that are allowed to monitor, review and appraise their own performance often run into similar difficulties – with the notable difference that these are typically entombed in classified document repositories. Certification of EMP hardness of various military and civilian systems in transparently-operated all-DoD test facilities is a *sine qua non* for programmatic integrity.

Mandated *Balance* Between Military- And Civilian-Directed Efforts. Encouragement by the Government of civilian builders and operators of EMP-vulnerable components and systems should be pervasively encouraged – *using all necessary means* – to design, build and operate EMP hardness into their systems. These means may range from building to GSA-issued EMP standards in order to provide equipment to any Government agency to GFE-type testing and certification of industrial equipments offered to the commercial market, akin to a UL seal-of-approval, and may include mandates to federal regulatory agencies, e.g., the FCC, which supervise and license EMP-vulnerable national infrastructure. In any case, the Congress should periodically re-visit and, as necessary, re-tool statutory language aimed at balanced hardening efforts in the military and civilian sectors, so that national security "front-doors" are not locked while "back-doors" are left wide open.

Obvious features of hardening of civilian infrastructure will surely include regularly hardness-tested "backbone" systems, e.g., for essential communication functions, and regularly exercised rapid-reconstitution ways-and-means, e.g., for restoring large-scale electrical power systems from hardened, distributed component stores. While largely civilian-sector in character, assuring that such minimum essential capabilities either reliably survive EMP attack or can be quickly re-constituted thereafter surely falls within the ambit of ". . .providing for the common defense".

Continuing Congressional Engagement. Accompanying all of the above is a need for continuing Congressional engagement with the DoD's best thinking and analysis, of the general character which is traditionally associated with

Congressional oversight proceedings which review mandated annual reporting and *ad hoc* certifications.

Congressional oversight with which I'm familiar in the strategic warfare area has been highly commendable in its peak intensity, its intellectual acumen and its cogency but, with all due respect, has been less-than-perfect in its regularity and follow-through. *Constancy and perseverance will be crucial in seeing Congressional mandates faithfully and efficiently translated into DoD programs and EMP defenses-in-being*, as cognizant Government officials-and-officers come and go. Commitment of highly capable, single-task Congressional staff members to such functions would have both symbolic and practical significance.

CONCLUSIONS. Electromagnetic pulse (EMP) is a "weapon of mass *hardware* destruction", even one instance of which can cripple the U.S. military machine and which also can lay waste to modern American civilization – without directly harming a single American. Technical means of defense against EMP exist which are of unquestioned technical feasibility and military effectiveness. ***Whether EMP defenses are financially and programmatically feasible to produce and deploy is for the Congress to determine – for corporate DoD seems to have elected to not face this esoteric but possibly historic issue.***

I thank the Committee once again for the opportunity to appear and comment on these matters of enduring significance for our Nation's security.

Information Submitted In Compliance With
Rule XI, Clause 2(g) of the House of Representatives

Dr. Lowell Wood is a Visiting Fellow at the Hoover Institution on War, Revolution and Peace at Stanford University, and a permanent staff member (currently on the Director's Technical Staff) of the Lawrence Livermore National Laboratory, operated under long-term contract (since 1953 through the present) by the University of California for the U.S. Department of Energy and its predecessor organizations, under Contract W-7405-eng-48. The Hoover Institution has received grants and contracts in support of its research activities from the Federal Government for many years, including the grant of its principal office building pursuant to an Act of Congress in 1974 memorializing President Herbert Hoover. The Livermore Laboratory, designated a National Laboratory by Act of Congress, has received total funding of the order of $1 billion (FY'96 $) annually for several decades from many agencies of the U.S. Government, with approximately 80% currently coming from the Department of Energy, approximately 10% coming from the Department of Defense and 10% from other Federal agencies.

Dr. Wood is unaware of any funding currently being received for any purpose by either the Hoover Institution or the Livermore Laboratory – or funding received during the past three years, or funding which is anticipated or being negotiated for – on the subject of his testimony. In particular, Dr. Wood has received or benefited from no such funding personally.

Dr. Wood is not representing the Hoover Institution or the Livermore Laboratory, or presenting positions of either of these organizations, in the course of his appearance before the Subcommittee. (To the best of his knowledge, neither of these organizations maintains any positions on any of the subject matter of his testimony.) Neither the Hoover Institution nor the Lawrence Livermore National Laboratory have contributed to the preparation of his testimony, which has been prepared by him as a private individual. Dr. Wood's testimony is offered as being solely his own, not necessarily representing the opinion of anyone – or anything – else.

CURRICULUM VITAE

Lowell L. Wood, Jr.

Home Address

2844 Royal Avenue
Simi Valley, CA 93065

Business Address

University of California
Lawrence Livermore National Laboratory
P.O. Box 808, L-278
Livermore, CA 94551-0808

Personal

31 August 1941, Santa Monica, California
United States Citizen; Married, 1 Child

Education

B.S., Chemistry and Math, U.C.L.A., 1962
Ph.D., Astrophysics, U.C.L.A., 1965
Dissertation title: "Hyperthermal Processes In The Solar Atmosphere"
Thesis advisor: Professor Willard F. Libby

Professional History

Visiting Fellow, Hoover Institution, Stanford University, 1994-present

Professional Staff Member, University of California, Lawrence Livermore National Laboratory (1972-present)
Director's Technical Staff and Associate Physics Department Head, (1972-1975)
Special Studies Group Leader, (1972-1985)
Special Studies Division Leader, (1985-1994)
Special Studies Program Leader, (1975-1994)
Advanced Technology Program Leader (1992-1994)
Director's Technical Staff (1996-present)

Assistant Professor, Department of Applied Science, University of California, Davis/Livermore (1966-1972, half-time)
Teaching and research in the applied physical sciences, with emphasis on computational physics and artificial intelligence, electrodynamics, biophysics and the conceptual foundations of technical innovation.

Consultant, University of California, Lawrence Livermore National Laboratory (1965-1972, part-time)
Research concerning the terminal phases of stellar life cycle, with emphasis on supernova radiative hydrodynamics; research into nuclear explosive-based national defense systems; studies of thermonuclear physics and technology culminating in technical proposals leading to the establishment of the national inertial confinement fusion program.

Assistant Research Geophysicist, U.C.L.A. (1965-1972, full- and half-time)
Research into nuclear reaction processes in the solar atmosphere, and terrestrial consequences thereof.

Awards and Honors

National Merit Scholarship, 1958
Phi Eta Sigma, 1960
Phi Beta Kappa, 1962
National Science Foundation Fellowship, 1962-63
NASA Predoctoral Traineeship, 1963-1965
Sigma Xi, 1965
Ernest O. Lawrence Prize (National Security), US Department of Energy, 1981
Abrahamson Award (Technical Leadership), American Defense Preparedness Association, 1991
Defenders of Freedom Award, Doctors for Disaster Preparedness, 1995
Golden Plate Award, American Academy of Achievement, 1997

Professional Memberships

American Association for the Advancement of Science
American Institute of Aeronautics and Astronautics
American Physical Society

Major Professional Interests

Advanced computing systems and related technologies; ultra-high power devices and systems; laser physics and technology; high-energy astrophysics, particularly stellar explosions; controlled fusion; biophysics; space systems; high national defense technology.

TESTIMONY

by Lieutenant General Robert L. Schweitzer, U.S. Army (Ret.)

Before the Joint Economic Committee of the U.S. Congress on June 17, 1997

radio frequency weapons and the infrastructure

I have been asked to talk to the overall subject of your hearing from a somewhat different perspective. Initially, it was to be from the one of what technology transfer means to a soldier. That part would have been fairly simple to address. Field soldiers are too busy to think much, if at all, about such transfers. That is, until they run across them on a battlefield where U.S. technology or materiel is being used against them. That happened in World War II when the residue of simpler technologies in the form of scrap metal was employed against us in the Pacific. It happened in Vietnam when some of our weaponry was obtained by our adversary. It happened again in Desert Storm when we ran across containers of U.S. materiel in the hands of Saddam Hussein's soldiers, materiel which had been channeled through Jordan. Then the fleeting reaction is one of anger and "why?" But soldiers--placed as they are since the time of the Roman legions in the sand, mud, rain and snow to fight decisive battles--are really too busy to brood much about such things. They are, however, grateful when Congress acts ahead of time to bar technology transfers, not only the simple ones of which I speak but the more serious, albeit subtle ones, which can affect the outcome of battles and wars.

Today there is a new class of radically new and important radio frequency weapons (RFW) which merits your attention as it emerges. And in this case, the horse is out of the barn. Transfers have occurred and are occurring. Equally true, however, is the fact that there are things that can be done to protect our nation, which is the underlying objective of today's hearing. Certainly one of these things is to recognize that export control documents, particularly the Militarily Critical Technologies List, need to be reviewed to determine if radio frequency technologies should be considered in the same careful way we do nuclear technologies. I respectfully suggest that this is the case; stronger controls are needed. One example is Reltron tubes which went to a friendly nation, one who sells products widely--sometimes to nations who do not like us. These tubes, which can be small or large, generate intense radio frequency pulses and can be used as RF weapons.

Before we go further I wish to state clearly for you and for the public record that I do not speak for the Department of Defense, for any military service or any government agency. I come before you only as one who has researched this area for the past year and is writing a White Paper on the subject, one which will be offered to DoD for their use and disposition.

Some of you may know about radio frequency weapons, where they came from, what they can do and what the implications are.

Although there are a number of groups and individuals concerned with this subject, I have found that somewhat paradoxically the word has not really gotten out in Washington itself. Despite the existence of a Presidential commission, an Infrastructure Protection Task Force, a Critical Infrastructure Working Group, an Information Warfare School at the National Defense University, and other working groups, to include divisions on the Joint Staff in the Pentagon, as well as a few very dedicated and brilliant mid-level people in DoD, a general understanding is lacking. This is true not only of RFW, but of their immediate threat to our DoD and national infrastructure. Indeed the term "infrastructure" is so amorphous that it lacks impact if not meaning. One of our first tasks will be to define what is the military and economic infrastructure and what in it is susceptible and vulnerable to RF weapons.

Some 90 to 100 references in 26 pages of the 70-page Quadrennial Defense Review speak to this new threat, but only to a discerning reader; the name for the class is not used. On the other hand, a recent search of the Internet found 2,400 to 2,800 references, while yet another, more thorough search found many tens of thousands of documents where the key words "radio frequency weapons" appear. Some very good people have written books and articles on the subject, the first revealing article known to me appeared in 1987 in the *Atlantic Monthly*, but for many reasons the knowledge is diffused. In the public sector the subject has yet to draw any real attention or concerted action.

To help set the stage, recognize with experts like a former NSA Director that we are the most vulnerable nation on earth to electronic warfare. This thought is echoed by a former CIA Deputy Director, and a former Deputy Attorney General who forecast that we will have an electronic Pearl Harbor if we do not accept a wake up call. Our vulnerability arises from the fact that we are the most advanced nation electronically and the greatest user of electricity in the world.

On the military side, as in the civilian sector, our current superiority is based on microelectronics. To prevail against us, an adversary must cripple, destroy or deny access to those same microelectronics. Can an adversary do so? Very likely, as this hearing will bring out. All of our military doctrine assumes extensive use of sophisticated electronics and communication systems to ensure information dominance and overwhelming battlefield success. As is the case with our civilian infrastructure and economy, our current dependence is large and will continue to grow. Because our battlefield success and the well being of our civilian economy--with which this committee is especially charged--are so dependent upon the effectiveness of our microelectronic-based systems, we should fully understand any technology that might be used to defeat our systems. This is particularly true of the newly emerging threat of radio frequency weapons. And even more importantly, we must develop countermeasures before such weapons are used against us.

Before going further, let me explain what these weapons are, where the Russian work has gone since 1949 and the applications of these weapons. If you are interested--as I believe you will be--you may wish to bring before you successive panels of our own leading scientists and experts. I have talked to many of them, heard them make presentations at conferences, and read their articles and books. I will be pleased to provide your staff with names of those who could provide this or other committees with a better understanding. I am also willing to assist in any way that might be helpful.

First of all, an RF weapon is one that uses intense pulses of RF energy to destroy ("burnout") or degrade ("upset") the electronics in a target. These weapons can be employed on a narrow beam over a long distance to a point target. They are also able to cover broad targets. They are categorized as high power microwave (HPM) weapons and ultra wide band (UWB) weapons.

The phrase non-nuclear electromagnetic pulse is sometimes used, because these weapons, which are indeed non-nuclear, project the same type of pulse we first learned of in conjunction with nuclear weapons. As a practical matter, a piece of electronic gear on the ground, in a vehicle, ship or plane does not really care whether it is hit by a nuclear magnetic pulse or a non-nuclear one. The effect is the same. It burns out the electronics. The same is true of the computers in this Senate office building, in industry, or on Wall Street.

There is another way these weapons can be delivered to a target, military or civilian. Here the term RF munitions, or RFM is used. Yet these too are properly called RF weapons. These small munitions contain high explosives that produce radio frequency energy as their primary kill mechanism. In the hands of the skilled Russian scientists, these munitions come as hand grenades, mortar rounds, or large artillery shells or missiles. Generally, they produce a short but very intense pulse. While not yet fully understood and with some uncertainties argued as to their capabilities, many scientists are convinced the weapons actually exist. Without making any claims as to what they can do, I offer the following list from open source FSU literature of some nine smaller RF munitions or weapons:

MAGNETOHYDRODYNAMIC GENERATOR FREQUENCY (MHDGF)
EXPLOSIVE MAGNETIC GENERATOR OF FREQUENCY (EMGF)
IMPLOSIVE MAGNETIC GENERATOR OF FREQUENCY (IMGF)
CYLINDRICAL SHOCK WAVE SOURCE (CSWS)
SPHERICAL SHOCK WAVE SOURCE (SSWS)
FERROMAGNETIC GENERATOR OF FREQUENCY (FMGF)
SUPERCONDUCTIVE FORMER OF MAGNETIC FIELD SHOCK WAVE (SFMFSW)
PIEZOELECTRIC GENERATOR OF FREQUENCY (PEGF)
SUPERCONDUCTING RING BURST GENERATOR (SCRBG)

Some of these weapons are said by the Russians to be now available as a hand grenade, a briefcase-like object, a mortar or artillery round.

Applications or potential targets (like those of the larger High Power Microwave weapons) would include all military computers, circuit boards, or chips, of any description, and include the following key components of our military and national infrastructure. They would have equal impact on civilian targets with the advantage less power would be required. Recall that the term "infrastructure" lacks clear meaning, but would include things like:

The national telecommunications systems
The national power grid
The national transportation system, to include especially the FAA but also such simple things as our traffic lights (with consequent gridlock)
The mass media
Oil and gas control and refining
Manufacturing processing, inventory control, shipment and tracking
Public works
Civil emergency service
Finance and banking systems (to include banks' ability to dispense cash)

This list of potentially vulnerable targets could and should be extended to include airplanes, ships, vehicles and the like. Of interest is the fact that we are doubly vulnerable because we are, and will remain, in an era of dual use of military and civilian systems. For example, 90% of our military communications now passes over public networks. If an electromagnetic pulse takes out the telephone systems, we are in deep double trouble because our military and non-military nets are virtually inseparable. It is almost equally impossible to distinguish between the U.S. national telecommunications network and the global one. What this means is that it is finally becoming possible to do what Sun Tzu wrote about 2000 years ago:

to conquer an enemy without fighting. The paradigm of war may well be changing. If you can take out the civilian economic infrastructure of a nation, then that nation in addition to not being able to function internally cannot deploy its military by air or sea, or supply them with any real effectiveness--if at all. Note: The paradigm has not yet changed; we need all our military forces.

Since 1949, the intense interest of the former Soviet Union in developing these weapons appears to have resulted from their recognition that they could not match the capability of Western electronics, and their belief that RFW have the potential to be effective against our sophisticated electronics. It is far less clear to me and to others why they are willing to transfer and proliferate the RF technologies they have developed so carefully and so well, but that they are clearly doing so. Should you wish, a future hearing by this or another committee could go into more detail.

President Yeltsin proposed to President Clinton a joint program for a "plasmoid defense" against ICBM's. While it is unclear to many scientists what President Yeltsin meant, such a defense, if attainable, might presumably set up a shield which would ionize the atmosphere and cause missiles to fail. Official Russian journals and publications show keen interest and provide many details about these weapons. A great amount of information is flowing continuously from three former Soviet Republics on their past and current programs.

We do know that the reduction in military spending by the FSU and many Western nations is prompting the defense industries of many countries to offer advanced weaponry to foreign customers to further their own research, development and industrial capabilities. This trend is almost certain to grow over the next 10 years.

From unclassified sources, we know that Russia, Ukraine, the United Kingdom, China, Australia and France are well ahead in this field, while Germany, Sweden, South Korea, Taiwan and Israel are emerging and have ample details of the Russian work and of the proceedings of more than 20 years of international conferences. Without going into any classified matters one may reasonably infer that the pariah nations have similar interests and some certainly have the financial resources to develop or procure RF weapons.

Russian and FSU information on RFW has been moving across borders for many years. International conferences beginning in 1949 have been a principal source of technology transfer. Scientists here and abroad have long exchanged papers, letters and, with increasing frequency, telephone calls.

The first Megagaussing Conference on the generation of high

power electromagnetic pulses took place in 1949 in Frascati, Italy. Russian scientists were key players in what has become a long series of presentations on the generation of electromagnetic power. Present at this and many subsequent conferences was the U.S. inventor of RF weapons, Dr. Max Fowler. His picture was placed over the center of the Moscow desk of one of his Russian counterparts who is a leader in the Russian development of the smaller version of these weapons. The latter is a key figure in the offer to sell RFW and RFM or their technologies to others.

EUROEM Conferences have been meeting (with name changes) for perhaps some 20 years at about two-year intervals. At the 1994 conference which was held in Bordeaux, France, the Russians made public many details of their long work in these weapons. Some of their papers deal with the strategy, tactics and techniques for the use of offensive RF weapons. Among nations participating were Iran and Iraq. At this conference the Russians talked about selling their technology and weapons to prospective buyers. I am told that subsequently a large number of nations have engaged them in some form of negotiations. Some of these "buyers" raise legitimate concerns.

The BEAMS conference (with name changes) has been meeting about every two years since 1975.

The EUROEM Conference met in Albuquerque in 1996; the BEAMS Conference met that same year, I believe in Prague. Attendance was open to all nations.

The next EUROEM and BEAMS conferences will meet in 1998 in the Middle East, two weeks apart in Tel Aviv and Haifa, respectively.

An International Pulse Power Conference held their tenth conference under that name in 1995, but has existed under other names for a longer period of time.

The International Particle Accelerator Conference has also met for more than 20 years.

The American Physical Society has a Plasma Physics Division which hosted (for more than 20 years) many conferences. Usually each one has several sessions on microwave generation.

And there are more. . .

Understanding the number, frequency and long standing nature of these conferences, you can perhaps better appreciate why I earlier said that the horse is out of the barn. Of interest, too, is the role of the United States in these conferences.

Indisputably, the U.S. is the scientific powerhouse of the world. We have initiated and hosted a number of these conferences, funded many of them to a significant degree, and played a prominent role at all. While we gain some information, our scientists will readily acknowledge the net advantage is always to other attendees.

Put another way, from a narrow technology transfer standpoint we have thus far lost more than we gained. However, even prior to the Internet no one could control the flow of ideas, especially among scientists. They like to talk especially about what they have achieved, and how they solve theoretical and practical problems. For decades our scientists have found their Russian counterparts to be brilliant, dedicated and creative. Personal relations are important and some have developed, but they are exceptional. For the most part the Russians have been ambiguous about their great work and often are mistrustful of Americans. We should move to change that by closer and warmer contacts as well as by efforts to enter into joint ventures--with all the travails that accompany such efforts. The Russians are intensely interested in our comments and some professional appreciation by their scientific peers of their decades of work on the offensive use of RF weapons. In my humble opinion they would prefer to work with our own distinguished scientists rather than others, but will sell their technology and products to others. I believe there is a real potential for joint ventures which could serve to constrain to some degree the proliferation of these weapons, especially to those who would do us harm.

To return to the earlier point about the need for better controls of technology transfer, consider these two counterpoints which illustrate the problem:

First: Although RF weapon components are on the Critical Technologies List, there are no up to date DoD guidelines or directives on this subject. An attempt to do so was made two years ago when little was known about the subject. As a consequence, decisions within the U.S. scientific community are becoming harder and dicier to make. There is a lack of clear policy guidance and direction.

Second: The first point is illustrated by the transfer of the Reltron microwave tubes. These tubes, which generate radio frequency power, cost a great deal of money to produce and test. The U.S. is the leader in high-power tubes and their associated power systems. but the market is really thin. Our tube industry has no current buyers here in the U.S. Without major contracts from foreign countries (France, the United Kingdom, Germany and Israel, among others), our tube industry will die. We will lose contact with real customers and become dependent on foreign hardware for our systems.

> Ultimately we will increase the difficulties that must be overcome to develop HPM applications for any future DoD use. Almost certainly we would know less--almost nothing--about what was going on in this area. For their part the Europeans and others would not cease to procure; they would simply undertake their own development. So our high power microwave scientific community told the State Department on balance to approve the transfer, which State did. Inevitably one consequence will be to advance the work of others and ultimately the production of RF devices to be used wherever and however by whomever. Note well, however: there is no guarantee that friendly countries will not sell the devices they produce to unfriendly, even hateful people.

It would also appear that there are other proliferation and transfer concerns of interest to this committee, simply because there is so much accurate how-to-do information in the open literature and on the Internet. Several countries have RFW programs and Russia says it has sold some technologies to these countries. At least one of these countries has acknowledged such a transfer. The crux of the difficulty in controlling these transfers is best illustrated by the fact that High Power Microwave weapons look like ordinary radars. With a dish or horn antenna, and a van with a power source, an RFW would look like a new, used or renovated radar. Used ones are offered for sale today in military surplus and commercial catalogs. Other catalogs offer for sale the components to put together lower power, but also very low cost items, that once assembled could be used effectively against the infrastructure.

Users of the new weapons can be criminals, individuals or organized gangs of narco or domestic terrorists--or a determined, organized, well-funded foreign adversary, either a group or nation who hates us.

The Russians, as noted, led with this work starting in 1949 with theory. By 1961, they were doing research, as documented in their numerous unclassified scientific articles. Experiments began in the seventies and proceeded to testing as described in their publications. Many of these weapons appeared in written descriptions, some photographs and diagrams in the nineties. Strategy, doctrine, tactics and techniques are all laid out in rather clear form. Please note all of this is unclassified information.

There is a legitimate question about the intelligence aspect of all of this. Our intelligence community largely proceeds on the operating principle followed in the Cold War: A threat is not validated until it is fielded. Well and good; hard evidence is essential.

But the question may fairly be asked: does that principle serve us well in the present day? Suppose we were to take a Russian or FSU-designed weapon, fabricate it in the U.S. and test it here. If the results were to meet the standards of performance and capabilities now claimed by the Russians, would we then have a validated threat? The answer to the capabilities may be forthcoming this month because at an unclassified level one of our national labs is doing just that. Another lab has purchased cheap, off the shelf components and will test its lower power device this month. Their engineers and I believe it will indeed work against infrastructure and light military targets.

There is a great deal of other corroborating evidence which at least argues for the existence--which is still disputed in some quarters--of these weapons. Evidence as to the capabilities of the weapons may be found in such recent statements as China's declared intention to purchase three RF weapons derived from the Russian technology. Another is the series of reliably reported discussions within the IRA of their intention to seek RF weapons for use against the London financial system in lieu of bombs and explosives. Consider, too, the recent statement by Sweden they have used these devices in experiments to stop cars at 100 yards, as well as their reported claim that RF weapons have been used against their financial institutions. A similar but much disputed statement has been reported by the *London Times* concerning British financial and banking institutions. The Los Angeles Police Department had done some successful work with vehicles in the interests of public safety and to halt fleeing suspects.

Advantages of the larger high power microwave RF weapons include:

- Low cost per engagement
- All weather
- Instantaneous engagement times
- Simplified pointing and tracking
- Possible to engage multiple targets
- Deep magazines simplified logistics (can "fire" or pulse as long as there is power in the generator)
- Non-lethal to humans when properly adjusted
- Well suited to covert operations because of lack of signature; deniability
- Not able to detect attacks; silent when used without explosive devices

The RFM offer many of the same advantages, offset only by the sound of the explosion that detonates them and produces the rise in pulse energy.

Unless we choose to be, we are not without courses of action. Some of these could be explored at a future hearing. Some preliminary thoughts are offered today:

We neither fully understand nor control this technology.

We have not begun to work on defenses , especially for our vulnerable infrastructure.

We need to first scope the problem, determine susceptibilities and vulnerabilities, then test.

All of this, to include any appropriate hardening of existing components, will take many years.

There are other courses of corrective action, but all will take time to acquire and apply.

The first step might well be to bring forward our real RF experts in DoD and the scientific community who know what needs to be done.

We need to go at this problem with a step-by-step sensible approach. No budget buster is proposed. Even if Congress had ready funds, a grandiose national solution is not the way to go.

We can start by scoping the problem and then by applying some of the same low-cost components that are now used in the ever expanding information technologies. Examples are surge-like protectors, plasma limiters, diodes, and metal covers. Parallel or redundant systems are another technique.

We are good at managing risks. We should no longer hesitate to reduce the impact of the threat, or to give our intelligence community the guidance to open up (some would say revise) their approach to this problem. Clearly the United States Congress will play a key role in whatever we do, or choose not to do, and our top leadership should focus on the longer term. But we should begin now in a sensible, modest way.

Three things we want to keep foremost in mind:

Do not throw a lot of money at this problem. Funds don't exist; the best solutions will have to be devised.

Do not tell DoD or the Services to take this out of their budgets. They are over stretched now and it would be wrong to tell them to pay for protection of the civilian infrastructure.

Do not continue to do what we have been doing and ignore the problem.

QUESTIONS AND ANSWERS SUBMITTED FOR THE RECORD

JULY 16, 1997

QUESTIONS SUBMITTED BY MR. WELDON

Mr. WELDON. What policies and standards guide those responsible for the acquisition of space systems with respect to protection of those systems from system generated EMP (SGEMP) and other nuclear effects?

Mr. KLINGER. System generated EMP (SGEMP) results from the interaction of X-rays with the body and electronics of a satellite. These effects depend upon the amount of X-rays, the energy spectrum of the X-rays and most especially, the satellite design. There are several sets of handbooks and computer codes developed by the Air Force and the Defense Special Weapons Agency that provide the initial, physics-based starting point for SGEMP analysis.

The Joint Chiefs of Staff (JCS) have established criteria that define the hardening requirements for all space systems based on their mission, their validated threat, and the natural space environment in which the satellite will operate. These requirements are articulated in a system's Operational Requirements Document (ORD), translated into its Technical Requirements Document (TRD), and tracked by the Defense Acquisition Board (DAB) throughout the system acquisition per the DoD series 5000 acquisition regulations.

The DoD Space Architect has recommended hardening approaches for all space systems. These guidelines will be used as the basis for requirements and capabilities tradeoffs during system acquisition.

Mr. WELDON. How costly is the protection of space systems from SGEMP? How high a priority is this protection for space systems?

Mr. KLINGER. Protecting satellites from system generated EMP (SGEMP) effects is primarily accomplished through designs that consider the entire electrical interconnectivity of the system. This cost is typically low if the requirement is included as part of the original design of the satellite and usually grouped with the cost of radiation hardened microelectronics parts. The Defense Special Weapons Agency has studied this issue and found that, typically, SGEMP and radiation hardened parts add 1 to 8% to the cost of a satellite, if incorporated in the original design. If appropriate protection measures are not included as part of the original design, then the cost to add protection measures can increase significantly. The increased cost would vary from satellite to satellite, depending on the satellite's complexity. Several factors determine the requirement for hardening a satellite; its natural operating environment and, most importantly, the validated threat to the satellite. Each system will have a different requirement for nuclear hardness.

Mr. WELDON. In establishing a space architecture, was the potential vulnerability to EMP effects of commercial satellites on which DoD now depends a serious consideration?

Mr. KLINGER. Yes. Assuming that the "EMP" above refers to High Altitude EMP (HEMP). Also, that the commercial satellites that DoD depends upon are primarily communications satellites at geosynchronous orbits. The potential vulnerability of these satellites is extremely low. However, that vulnerability was considered in establishing a space architecture.

Mr. WELDON. How vulnerable to EMP are the ground facilities needed to operate our space systems?

Mr. KLINGER. Vulnerability to EMP is a consideration in the overall design, construction and operation of satellite system ground stations. Detailed analysis and status is best addressed in a closed session.

Mr. WELDON. A recent "Army After Next" war game started with a concerted attack by "Red" forces that knocked out our satellites with EMP and other nuclear effects. Maj. Gen. Robert Scales from the Training and Doctrine Command, who runs the Army After Next program, now argues that all our satellites—even commercial ones—must have self protection capabilities. What is your assessment of the realism of the exercise? Do you agree with the conclusion drawn by General Scales?

Mr. KLINGER. The Army After Next (AAN) Program provides the Army leadership strategic insights based on operational analysis of war in 2020. It is premature to draw any final conclusions from AAN war game results. The Winter War Game (WWG) in February 1997 was notable for a space campaign preceding the engagement of opposing ground forces. That space campaign included many hypothetical weapons systems and culminated in a nuclear attack on satellites. This direct nuclear attack did not utilize EMP as a kill mechanism.

Providing all satellites with self-protection is neither affordable nor necessary. System-level (architectural) trades have also shown that it is more important to protect the function, rather than protecting the individual assets that provide the function. The degree of protection for a space system is determined by the system's planned use, consequence of loss, potential threat to the system and cost.

Mr. WELDON. How routinely do we test for EMP vulnerability?

Dr. ULLRICH. Not as often as we used to test. In fact, very few systems are now tested because there are fewer new systems and requirements for EMP survivability have been reduced in recent years. Generally, extensive EMP testing is restricted to new systems and those with major retrofits. Periodic hardness maintenance and hardness surveillance tests should take place to preserve the hardness of fielded systems, but the amount of this testing varies from system to system.

Mr. WELDON. How good is the fidelity of the testing?

Dr. ULLRICH. The test fidelity for smaller, self-contained systems, such as missiles, small aircraft, tanks, and mobile vans, can be quite good. High level EMP test facilities, or "simulators," are currently available in a number of locations around the country. A simulator of this type can illuminate a modest exposure area with a "threat level" electromagnetic field that very closely approximates a nuclear EMP pulse.

Distributed systems (such as the national power grid, or the commercial communications network), and systems that are attached to long conductors (such as a ground facility connected to power and telephone lines, or a train on a track), cannot be moved to a threat level simulator. The test fidelity is not as good for these systems because the testing has to be done on-site with transportable EMP simulators. Such simulators generally have lower simulation fidelity than threat level simulators. Additionally, the entire system cannot be illuminated simultaneously, so the system is tested a section at a time. Since it is usually not practical to test the entire system, analysis must be used to determine the overall system response.

Mr. WELDON. How predictable are the consequences of EMP on large systems?

Dr. ULLRICH. Predictions of the consequences of EMP that are based only on analytical methods are less accurate. This applies to both small and large systems. Recent DSWA research demonstrated that the complexity of modern electronic and electrical systems resulted in large errors in analytic predictions of the EMP effects on systems. The conclusion of the research was that high level testing of the full system is required to arrive at more accurate assessments of system response.

Mr. WELDON. Once tested and deemed secure from EMP effects, how often are facilities retested?

Dr. ULLRICH. Recommendations for revalidation testing have been from five to seven years or whenever there is a major modification of the system. Our experience has shown that hardness maintenance and hardness surveillance programs are not consistently implemented.

Mr. WELDON. How are large systems tested for EMP resistance—for example, how is an ICBM launch facility or B–2 bomber tested to assure that it can withstand EMP effects?

Dr. ULLRICH. ICBM launch facilities can be tested by using a combination of field illumination simulators that apply EMP-like fields over the facility and current injection simulators to induce large currents on the external electrical cables and penetrations.

Large aircraft, such as the B–52 and the B–1 bombers, have been tested using the TRESTLE, HPD and VPD simulators located at Kirtland AFB, NM. The aircraft is placed within the exposure area and the entire aircraft is illuminated with EMP-like fields with their engines running and mission equipment operating. Such tests can also be combined with current injection tests to drive cables and the skin of the aircraft.

Mr. WELDON. Is component level testing capable of providing high levels of confidence that our strategic systems are not vulnerable to EMP?

Dr. ULLRICH. Component level testing alone is not capable of providing high levels of confidence that our strategic systems are not vulnerable to EMP. Successful EMP test programs have relied on a series of tests to achieve high confidence. These tests include component burnout tests, functioning subsystem current injection tests and high level, EMP-like field illumination of an entire functioning system. The final confidence is gained by the combined results of all the tests, not just a single type of test.

Mr. WELDON. How robust are our EMP testing facilities? What EMP testing facilities are currently operational? Are they in good condition? Which have recently closed and why?

Dr. ULLRICH. Existing EMP facilities are generally physically robust. The most important component is the pulse power source. The pulsers require periodic maintenance to avoid serious degradation.

With the end of the Cold War and the accompanying budget constraints, many of the large, system-level EMP simulators have been mothballed or decommissioned. The table provides a brief status of the major DoD facilities. A brief description of the simulators follows:

HEMP Simulator	Location	Status
ARES [1] [DSWA]	KAFB, NM	Operational, recently upgraded.
FEMPS [2] [DSWA]	San Leandro, CA	Operational.
HPD, VPD [3] [Army]	KAFB, NM	Operational, planned for mothballing.
HPD, VPD [4] [Navy]	Patuxent River NAS, MD	Operational.
Empress I, II [4] [Navy]		Decommissioned.
Trestle [5] [Army]	KAFB, NM	Mothballed.

[1] ARES—A high-level, high altitude EMP (HEMP) simulator that can illuminate missiles and mobile ground equipment with threat level fields. Operated by DSWA at Kirtland AFB, NM.

[2] FEMPS—A smaller, high-level HEMP simulator that can illuminate racks of equipment and other smaller systems. Operated by DSWA.

[3] HPD, VPD—Horizontally Polarized and Vertically Polarized Dipoles that were originally built to provide high level field illumination of aircraft. They can also be used to illuminate mobile ground systems. Those at Kirtland AFB were originally built by the Air Force but have since been turned over to the Army and may be mothballed. Those at Patuxent River NAS are still used by the Navy to test aircraft.

[4] Empress I and II.—Surface ship HEMP simulators built by the Navy. EMPRESS I was on land and illuminated ships in the Chesapeake Bay. EMPRESS II was on a barge and was used to illuminate threat level fields on surface ships at sea. Neither simulator is now operational. EMPRESS II has been dismantled.

[5] Trestle—The Air Force built the TRESTLE facility to EMP test aircraft in an inflight condition. This facility has been mothballed.

In addition, DSWA has a Continuous Wave (CW) Measurement System that consists of low level CW field illumination and high level current injection. This system is transportable and allows testing of ground facilities (fixed and mobile) such as satellite communication terminals.

Mr. WELDON. Have tests been conducted that indicate that US ICBM launch facilities are vulnerable to EMP effects?

Dr. ULLRICH. Any comment on actual test results of specific U.S. systems would be classified; this question should be referred to the Air Force.

Mr. WELDON. Are there significant differences between Russian doctrine and Soviet doctrine on the use of EMP to disrupt or disable U.S. strategic and other military systems?

INTELLIGENCE COMMUNITY WITNESSES. The information referred to is classified and is retained in the committee files.

Mr. WELDON. Are there any other nations with nuclear forces now capable, or potentially capable, of generating an EMP threat to U.S. strategic forces or U.S. strategic command and control?

INTELLIGENCE COMMUNITY WITNESSES. The information referred to is classified and is retained in the committee files.

Mr. WELDON. Do we have any indication that these other nations plan to use nuclear weapons specifically to generate high altitude EMP?

INTELLIGENCE COMMUNITY WITNESSES. The information referred to is classified and is retained in the committee files.

Mr. WELDON. How mature is RF weapons technology? What nations are developing this technology? Does the intelligence community have any projections about what nations might acquire such weapons?

INTELLIGENCE COMMUNITY WITNESSES. The information referred to is classified and is retained in the committee files.

Mr. WELDON. Is the intelligence community aware of any potential terrorist threats to use RF weapons against key commercial targets?

INTELLIGENCE COMMUNITY WITNESSES. The information referred to is classified and is retained in the committee files.

QUESTIONS SUBMITTED BY MR. BARTLETT

Mr. BARTLETT. To what intensity level?

Mr. ULLRICH. The actual maximum intensity (peak electric field) level varies from simulator to simulator and within each simulator it varies based on the test object's location and extent. A nominal unclassified intensity that most threat-level simulators can generate is 50,000 Volts/meter.

O